GROWING CAMELLIAS

'Governor E. Warren'

Margaret Tapley

CASSELL

Acknowledgments

I wish to thank my family for their encouragement; Walter Logerman for his invaluable help again with all things to do with word processing; Michael Goodwin for computer tuition; the New Zealand Camellia Society, especially the Canterbury Branch, for assistance and encouragement; Woodland, Westhurst and Portstone nurseries and J Elliot's Nursery, Amberley, for information and help with photographs; also especially Charles Harrison for photographic material, and Neville Haydon in Auckland for his great knowledge of new varieties; Les and Gordon Matheson and Barry and Joan Jackson for advice and checking of varieties; Barry Sligh of Taunton Gardens & Nursery, Dame Ann Ballin and Vonnie Cave for help with photography. Gina Williamson has been generous in allowing photographs and sharing her beautiful garden. Without you all the book would not have come to fruition. Thank you again my friends.

While care has been taken to get the colour in the photographs as accurate as possible, colours can vary from flower to flower within the same variety.

Cassell Publishers Limited
Wellington House, 125 Strand
London WC2R 0BB

First published in Great Britain 1996
in association with David Bateman Limited,
Tarndale Grove, Albany Business Park, Bush Road,
Albany, North Shore City, Auckland, New Zealand

Distributed in the United States by Sterling Publishing Co. Inc, 387 Park Avenue South, New York, NY 10016, USA

British Library Cataloguing in Publication Data
A Catalogue record for this book is available from the British Library

ISBN 0-304-34838-4

Printed in Hong Kong by Colorcraft Ltd

Front cover: 'Dr Clifford Parks'
Back cover: 'Berenice Boddy'

'Anticipation'

Contents

INTRODUCTION

'Forest Green'

THE *Camellia* has had a long and illustrious history, the first recorded mention being about 1725 BC when a Chinese emperor declared that tea (an infusion of the leaves of *Camellia sinensis*) was his favourite drink. Many countries have printed postage stamps depicting the camellia. It has been made a religious symbol, and been planted in temples and graveyards. It has been immortalised in art and fable, drama, literature and the modern cinema.

The plant has provided oil for cooking, the manufacture of cosmetics, and found countless uses in industry. Popular both outdoors and as a greenhouse plant in the 1800s, it supported a thriving cut-flower industry in the South of France. It has been carried on tea clippers from Asia to Europe and intrepid plant-hunters have risked their lives (and sometimes lost them) to search for new species and varieties and to bring back seed and propagating material from far-flung countries of Asia for the pleasure of the Western world. More than 200 different species have been found in the wild.

Colonists carried camellia plants and seed from the 'Old World' of Europe to the pioneer gardens of the 'New World' —America, Canada, South Africa, Australia and New Zealand. Camellia societies, comprised of experts and enthusiasts, have been formed in separate countries all over the world, to study the genus and to pool their knowledge on all aspects of camellia culture, propagation and hybridising. Bees must have been the original hybridisers, but hybridising by people to produce more diverse and original camellias has been

'Australis'

going on for centuries — records say as early as the tenth century.

The most widely grown ornamental camellia variety, and that which first reached the Western world in the 1700s, was *C. japonica*. Because of its great diversity of form and colour, in the flowers as well as the bush shape, it became very popular throughout Great Britain and Europe in the early 1800s. From the plant-hunters, more species became available for hybridising during the nineteenth century and during the early twentieth century a big breakthrough came when the species *C. saluenensis* was sent back from Asia, for hybridisers found that it contributed much-needed hardiness, longer flowering time, and a self-grooming habit to the hybrids it produced. Later, as relations between Asia and the West improved after World War II, and Chinese botanists discovered newer species with quite different characteristics from the more well-known *C. japonica,* they generously shared plant material and knowledge with hybridisers in Great Britain, Europe, America, Australia and New Zealand. Thus the nature of the genus, as offered from nurseries to the home gardener, changed, becoming more varied and versatile.

The introduction of the winter-flowering variety, *C. sasanqua,* extended the flowering season. Other new species added perfume, more closely set flowers, smaller flower size and smaller leaf size. New shapes were added to the form of the camellia bush by hybrids crossed from other new species. Now available are dwarf, weeping and ground-covering camellias, also varieties suitable for bonsai, hanging baskets and container growing. There are upright forms, like cypresses, for formal planting on either side of steps or doorways, and there are hedging varieties which thicken up well and make wonderful interior divisions, or exterior hedges, for gardens. The large, open-growing, tree-like *C. reticulata* species from the province of Yunnan, in Southern China, are more sun-tolerant than *C. japonica* and make magnificent specimen trees, clothed as they are in spring with huge butterfly-like blooms of an almost iridescent quality. Other new hybrids are suitable for growing on walls and fences in the form of an espalier, and some of the slender, whippy-stemmed varieties will even train on a pillar or archway to give early spring flowers and make an evergreen foil for an adjacent, later-flowering deciduous clematis or rose.

Extension of the colour range of camellias has been another aim of hybridists this century. At present there is a spectrum from white through all shades of pink and red to an almost black-red. Interest has been heightened, however, by the discovery in China of the yellow-flowered species *C. chrysantha,* and of several other species with yellow blooms. So far hybridising results have been disappointing, but work is continuing in this area in the hope of extending the colour range into golds, yellows and apricot-oranges. Flowers have been produced with lavender and purple overtones, but so far no pure blue-coloured flower has been achieved by hybridists.

The camellia as a garden subject has become more adaptable and varied and so more useful, giving a solid evergreen background to the urban courtyard as well as the larger country landscape design.

The exchange of visits and information, and the continuing cooperation between camellia societies throughout the world, are confirming the present status of the camellia as a versatile garden plant. We can assuredly look forward to more interesting new hybrids in the years ahead, which will consolidate the present upsurge of interest in the genus and confirm its growing popularity.

Chapter 1

DESCRIPTION OF THE GENUS AND HISTORY TO 1900

'Berenice Boddy'

THE *Camellia* belongs to the botanical family Theaceae, which, with the approximately 30 other genera within this family, is a member of the tribe Gordonieae. The common characteristic shared by all members of this tribe is the formation of seeds within a capsule. It is useful to note some of the camellia relatives which belong to the family Theaceae, and which generally enjoy similar growing conditions, thus making interesting companion plants.

The attractive and useful gordonias, found in both Asia and America, are very similar to the camellias and have even been successfully crossed with them. *Gordonia axillaris,* from Asia, is a small tree which has shiny green leaves and crepey white flowers which appear in late summer, making it most useful to succeed the camellias, which bloom in winter, spring and early summer.

Stewartias, which hail from the same continent as gordonias, are another useful relative. *Stewartia pseudocamellia* is a small tree whose leaves turn purple in the autumn, making a good foliage contrast to the evergreen camellia. The white flowers are 6 cm across and very like camellia blooms, with orange anthers but a furry reverse to the petals.

Cleyera japonica provides two varieties with interesting foliage. The form the Japanese call 'Sakakia' has tiny, fragrant, almost invisible flowers which are creamy-yellow with brown stamens. A cultivar, 'Tricolor', has green leaves, variegated yellow and pink.

Schima superba (syn. *S. wallichii*) is a large flowering tree with fragrant white

'Desire' is just one example of the many flower forms of *C. japonica* hybrids.

Like most reticulatas, 'Valley M. Knudsen' is taller growing and has larger blooms than many japonicas.

flowers flushed red in bud, followed by colourful fruits, but it is suitable only for tropical areas. *Eurya, Franklinia, Ternstroemia,* and *Tutcheria* are other tropical evergreen relatives.

Species

The genus camellia has been divided into species according to floral and leaf characteristics. Nearly 200 species of camellia have been classified into four subgenera and 19 sections.

New species are still being discovered in China, and shared with botanists in the Western world, to add to a genetic base of material being used for hybridising in the search for different attributes of colour, scent, leaf and bush shape in modern hybrids.

Some interesting species and the unique features which they transmit to their hybrids

Camellia sinensis (var. *sinensis*), the tea plant, is a bushy plant with very small white flowers and long, narrow, crinkled leaves. From this plant comes the unfermented green tea widely used in Japan. *C. sinensis* var. *assamica* and its hybrids pro-

vide most of the tea drunk in the Western world.

The earliest in cultivation and the most commonly grown of the ornamental species is *C. japonica*. It has strong growth, glossy green foliage and varied blooms of excellent texture — all desirable traits, which make it a popular choice for hybridising, and more than 20,000 cultivars have been produced from this species. The ability to produce a great diversity of flower form, size and colour, and also of bush form and leaf shape, in its hybrids is its chief attribute. It can be damaged by severe frosts and does not like full exposure to strong sunlight. Higo camellias are not a distinct species but a form of *C. japonica* which produces a profusion of prominent stamens. They can be single to semi-double in form.

C. reticulata comes from the Yunnan province of southern China. It is taller growing and more open in structure than *C. japonica,* with leaves which are longer and more pointed. Flowers are large and exotic, and it tends to produce larger blooms in its hybrids.

C. sasanqua flowers from autumn or the beginning of winter. It originates in Japan

The *C. lutchuensis* hybrid 'Quintessence.'

ing period. It readily imparts these assets to its hybrids. It is one parent of the famous *williamsii* hybrids. The semi-double flowers vary from white to deep rose.

C. tsaii is a miniature, white-flowered species with long, pointed, narrow leaves, which it tends to reproduce in its hybrids. It is often coupled with *C. rosiflora,* a popular pink-flowered miniature which makes a superb container plant. Growing into a small tree with a spreading habit and weeping branches, it often has variegated foliage. China was its original home.

C. lutchuensis grows in its native state only in the southern Japanese islands, including Okinawa. It is the most fragrant of the natural camellia species, growing to 3 m. It has small pointed leaves which have the added bonus of being red-tipped when young. With its pink-stained-white, sweetly perfumed flowers it makes a charming container specimen. It is much used in breeding programmes.

C. pitardii comes originally from southern China and is found in several varieties. They are small trees, up to 7 m. The dainty single flowers appear at mid season and vary from rose-pink to white. This variety has been much used in hybridising recently because of its hardiness.

C. fraterna, another species which is mildly fragrant, comes from central China and is also being used widely in breeding programmes. It may reach 5 m in height when growing in the wild. Its white flowers are sometimes tinged with lilac and it blooms at mid season. It has unusual leaves, elliptic in shape with pointed ends and bearing black-touched serrations on the base greyish-green colour. It is popular as a tub specimen.

C. transnokoensis, a native of Taiwan, has become popular as a garden plant because of its delicate fragrance and exquisite white flowers, which appear at mid season. A

or the Liu Kiu Islands. Some sasanqua hybrids have a distinctive musky perfume. Flower colour is white, through the pink shades to red. Sasanquas have a more spreading, willowy habit of growth than japonicas, are often used as espaliers, and can be trained to cover walls and fences. They are more sun-tolerant than japonicas and this attribute, and their early blooming habit, are often imparted to their hybrids. *C. hiemalis* and *C. vernalis* are often found listed with *C. sasanqua* in garden centres and are very closely allied. *C. hiemalis* flowers in winter and *C. vernalis* in early spring.

C. saluenensis is cold-resistant, floriferous, self-grooming and has a long flower-

C. transnokoensis

superb hanging-basket plant, it is also excellent in a tub.

C. granthamiana was found growing on a hillside in Hong Kong's New Territories as recently as 1955. It has very large (16 cm across) white flowers with eight petals, opening early in the season from buds which appear brown and dead. The flowers are distinguished by beautiful drooping golden stamens.

C. forrestii originated in Yunnan and the Tonkin area. It has very small leaves and white flowers which are distinctly fragrant.

C. kissi grows naturally from northeast India to southern China and the island of Hainan. The small, white, often fragrant flowers appear mid to late season and are single.

C. chrysantha has excited hybridisers all over the world. It promises a new colour in the camellia's range as it is yellowed-flow-

ered. So far there has been little success in producing new crosses but a further 22 species with yellow flowers have been found in China, giving strong hope that a suitable one for hybridising will become available.

This is but a brief outline of some of the many interesting species being grown as attractive garden and container plants and being used in breeding programmes to increase the range and scope of the genus, especially in perfume, extended colour range, growth habit and flower size.

Classifying camellias

The camellia is a versatile and indispensible asset for the creative gardener, either amateur or professional, particularly because of its great diversity of form, size, growth habit, leaf colour and size, and flower form, colour and size. To cope with such variety there are some guidelines which help to describe and classify camellias, whether we are choosing them for the garden, entering a specific class in a camellia show, or buying a gift for a friend.

Flower size

This can vary from the smallest at 1 cm, e.g., *C. forrestii,* to the largest at 20 cm, e.g.,

C. granthamiana

'Pink Pagoda' (above right) has an imbricated flower form, in contrast to the fimbriated petals of 'Fimbriata' (above).

'Jean Pursel'. When applied to camellias the terms 'miniature', 'small', 'medium', 'large' and 'very large' refer to the flower size and not to the height of the bush. 'Spring Festival', for instance, is a tall, upright, slender-growing variety in shape, which is most useful to the landscaper requiring a 2-m punctuation mark for the back of a border, but the pink double flowers with which it is

'Spring Festival' is a tall, upright variety with miniature flowers.

thickly smothered in mid to late season are *miniature* in size.

The standard accepted ranges for these categories of flower size are:

Miniature	less than 6.5 cm (2.5 inches)
Small	6.5–8 cm (2.5–3 inches)
Medium	8–10 cm (3–4 inches)
Large	10–12.5 cm (4–5 inches)
Very large	12.5 cm (5 inches)

A further category for **tiny** or very small blooms has been used by some nurseries.

Flower forms

Camellia flowers are classified by the number and arrangement of petals.

Imbricated — petals which are tapered to a point and overlap each other closely.
Fimbriated — fringed.
Petaloids — stamens on which the anthers have become flat and petal-like.
Rabbit ears — small, narrow, often twisted, upright petals among the stamens or other petals, and a stage larger than petaloids.

Official definitions

Single — one row of not more than eight petals and with conspicuous stamens.

Semi-double — two or more rows of petals and with conspicuous stamens.

Anemone form — one or more rows of large outer petals around a central convex mass of intermingled petaloids and stamens.

Peony form — the flower is a convex mass of intermingled petals and petaloids either with or without stamens. Some authorities prefer to divide the peony form into a full peony form and a loose peony form, the word 'loose' in this connection meaning petals which are not close or compact in arrangement.

Rose-form double — imbricated petals, showing stamens in a concave centre when fully open.

Formal double — fully imbricated, having many rows of petals never showing stamens.

Sports or mutants

Sports or mutants sometimes develop spon-

Left: 'Wilamina' is a small formal double.

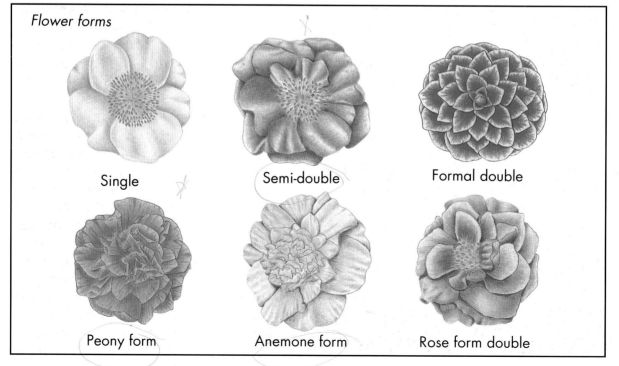

Flower forms

Single

Semi-double

Formal double

Peony form

Anemone form

Rose form double

Above: 'Leonora Novick' has a large to very large loose peony form.

Below: The semi-double flower of 'Shot Silk'.

A sport, or mutation, of 'Countess of Orkney'.

taneously without the intervention of people or cross-pollination by bees. New features develop which have quite different characteristics from the original variety on some parts of a plant. This 'sporting' phenomenon occurs most often with *C. japonica* varieties, rarely with those of *C. sasanqua* varieties and not at all with varieties of *C. reticulata*. The reasons for sporting are too complicated botanically to be dealt with in detail here, but lie in the evolution and heredity mechanisms of the species.

When mutations appear on the growing part of a plant they are commonly called bud-sports, and it is by vegetative propagation of these bud-sports that new varieties come into being. The most common change is flower colour and it is the most easily noticeable. Other possible changes are in flower form, changes in petal edge, leaf edge, leaf form or leaf variegation. An example of a famous sporting variety is 'Elegans', bred in England by Alfred Chandler in 1831. The description then was 'having free growth, the flowers being a delicate rose'. Famous sports which have since been produced from this variety are 'C. M. Wilson', a light pink, edged white, which sported to 'Hawaii', a pale pink, fimbriated, paeony form, and 'Elegans Splendour' which is light pink, edged white, with deep petal serrations. 'Elegans Champagne' belongs in the same family and is white with creamy-yellow central petaloids and has a wonderful 'champagne' look. A most attractive new sport was developed from 'Hawaii' by New Zealander Tony Armstrong and his late wife, Lorna. It is white, fimbriated, and beautiful, and has been named 'Hawaiian Bride'.

Hybrid camellias

There is now such a huge variety of camellia cultivars in existence that I will organise their description into broad groups of hybrids with similar characteristics.

1. C. japonica hybrids These are the most numerous and diverse in flower form, accounting for more than three-quarters of the camellia cultivars grown in the world. They are the traditional, broad-leaved camellias, the earliest to be introduced in the Western world. They thrive best in more temperate conditions nearer the coast and in general they are more sensitive to very hard frosts and severe winters with frequent snows. English writers have stated that flower size diminishes as they are grown in more northern regions of Scotland, and it has been noted that flowers are larger on cultivars grown in milder climates.

C. japonica hybrids are dense, slow growers or very fast, open growers so that there is no typical C. japonica cultivar shape. They can be low and spreading, fine, rounded bushes or tall and slender, open or compact shrubs. They may vary from almost prostrate forms to more or less fastigate, upright forms. With this variety of bush shape comes a similar variety of flower shape, a wide range of colours and

a wide range of flowering times during the spring and early summer. This variety makes them very versatile and deservedly popular as a landscape plant. Older varieties are still being grown and treasured and newer, more modern ones have been developed to excite the grower.

2. C. sasanqua hybrids (and including C. hiemalis and C. vernalis hybrids) These three are often grouped together in catalogues, for they all bloom during the autumn and winter, their history in Japan is at least as old as the japonicas, and there has been much hybridising within the three species. The flowers of all three groups shatter badly, so they are not so good as cut flowers and they leave a carpet of fallen petals around the bush. They are more sun-resistant than japonicas, making them suitable for hedging, espalier, container growing, topiary, groundcover and bonsai. They have a long history in Japan as garden plants, going back as far as the fourteenth century; 400-year-old trees can be found in temple gardens. The colour range is not as wide as that of the japonicas, there being few really bright reds and fewer variegations, but there are many two-toned 'sweet pea' types with delicate colours bordering the petals, and fluted and ruffled petals. Flower forms are not as varied as the japonicas, but there are wonderfully decorative stamens to be found. Some are widely flared, as in the Higo camellias. Many of these hybrids have a distinct fragrance, not always sweet, sometimes quite earthy, but most enjoyable on the sharp morning air.

3. Hybrids of C. reticulata This species comes from the province of Yunnan (the 'Land of Eternal Spring' or 'Kingdom of the Flowers'). Scattered stands, some of them hundreds of years old, exist, and the species also can be found as isolated specimens up to 20 m in height. The plant is tall,

Sasanqua hybrids such as 'Shishi Gashira' have a long history in Japanese gardens.

open and tree-like, with huge blooms of unusual beauty displaying iridescent colouring over a long flowering season. *C. reticulata* hybrids require a lighter, better drained soil than the japonicas and they do not enjoy heavy shade. They should be placed so they receive sun for some part of the day at least. They make splendid specimen trees when grown to their full magnificent width and breadth. Their looser, more open growth makes them not as suitable for hedging as the japonicas and sasanquas. Judicious pruning ensures their magnificent flowers are seen to their best advantage.

4. Other hybrids After World War I many plant-hunting expeditions sent new species of both rhododendrons and camellias from China and the Himalayas back to Europe.

Among the camellias was *C. saluenensis,* which produced, when hybridised with *C. japonica* cultivars, the famous *williamsii* hybrids named after Mr J.C. Williams of Cornwall, who made the first cross, adding hardiness, a long flowering season and the habit of self-grooming to the hitherto rather tender disposition of the japonicas. A wave of beautiful hybrids followed.

After this early success, hybridists began crossing established species and cultivars with the recently discovered *C. cuspidata, C. fraterna, C. granthamiana, C. heterophylla, C. lutchuensis, C. oleifera, C. pitardii, C. rosiflora, C. tsaii* and many other new and interesting species, with different, perhaps weeping, forms, smaller leaves or perfume. The camellia now acquired more versatility. There were scented varieties, there were weeping, slender-branched bushes, there were small leaves and miniature flowers like apple blossoms, there

Smaller weeping varieties are ideal in hanging baskets.

were groundcover dwarfs and plants with pendant leaves and small scented flowers which could be used very well in containers and hanging baskets. The camellia could now be used in small urban gardens as well as larger country ones.

Early history of camellias in cultivation

The geographical regions where camellias grow in their native state are China, Japan and Indo-China, and it is in these areas of ancient cultural development that the breeding, selection and cultivation of the camellia has gone on and been recorded in written historical documents for well over a 1000 years before the genus was known to the Western world.

In the art of China, even before the Tang Dynasty (AD 618–907), double and semi-double forms had been depicted on scrolls, on inlaid boxes and painted on porcelain. *C. reticulata* was shown being grown in gardens and cultivated in Buddhist temples. Other uses for this versatile plant were as a source of oil for food, cosmetics, and industrial purposes, and for the manufacture of charcoal for fuel. The most widespread use of all was as the beverage now called 'tea' by the Western world.

In contrast to the formalised, work-intensive and carefully trimmed and tended gardens of Egypt, and much later of Europe, the ancient Chinese philosophy of gardening saw Man as an integral part of his environment, not as a dominating and controlling force. Their magnificent landscape gardens, as portrayed in the paintings of early artists, tell us that they were designed to enhance the ultimate beauty of the natural world rather than force unnatural designs upon it. Camellia trees hundreds of years old can be found in monastery and private gardens. In China today, at the Kunming Institute of Botany, is housed a camellia collection of great note, and the scientists who work there are world leaders in their field.

In Japan, also, the camellia has played a significant part in the developing culture. The traditional 'tea ceremony' of Japan is remembered as a highlight by many visitors from Western countries, and many communities have selected the camellia as a symbolic flower. It has been depicted in the art of the country in textiles and ceramics, and has been the subject of paintings and decorations. The Japanese arts of bonsai and ikebana have both leaned heavily on the camellia for inspiration.

In the Shinto religion, with many adherents in Japan, the tsubaki, the native camellia of Japan or 'the tree with shining leaves', was an important symbol. The belief was widely held that the flowers were inhabited by Shinto gods in spirit form and camellias were found growing in many temple gardens and graveyards. Many of these older varieties have been propagated and reintroduced to today's growers.

Trade was closed between Japan and Europe during the eighteenth and nineteenth centuries, but a treaty in 1849 ended this isolation, and Japan's rich horticultural heritage could be shared with the rest of the world. At this time many Japanese

'Captain Rawes' was named after the captain of the trading ship that transported this reticulata to Britain last century.

camellia names were changed to European ones, causing much confusion for future scholars of the genus.

The flourishing tea trade which developed between China and Europe in the seventeenth century was the means by which the West first gained access to the camellia. The demand for the tea plant (*C. sinensis*) led to interest in ornamental camellias, as *C. japonica* was sometimes mistakenly sent to Europe instead of *C. sinensis*. The fascination which camellias had aroused in ancient China and Japan became apparent in Europe, and later spread to the New World.

There are many intriguing stories of how the camellia came from Asia to Europe and the New World. Plant-hunters were sent out to China, financed either by large private gardens in Britain or by horticultural societies. Itinerant missionaries, many from the Jesuit order, played an important part in both discovering new plant material and sending it safely out of Asia. The daring captains of the Dutch, English and Portugese East India Companies saw that they were safely transported to Europe, although there were frequent losses due to the hazardous sailing conditions of the time. The doctors aboard the ships were often the botanical recorders too, as medicine and botany went hand in hand at this time, and several of these men were instrumental in recognising and transporting new species of camellia to Europe.

In 1735 the genus was named 'camellia' by the Swedish botanist Linnaeus, who wanted to honour the work of a German missionary-botanist, George Joseph Kamel. Unfortunately, although Kamel had done

'Elegans Supreme'

some very valuable work on the flora of the Philippines, he knew nothing of camellias. The exact date when the first camellia arrived in Europe is uncertain, but we do know that in 1739 Lord Petre, who was a noted botanist and supporter of horticulture, had at Thorndon Hall in Essex, a semi-double red camellia growing in his glasshouses. This had probably come from China, as he called it 'the Chinese rose'. We know from records that a beautiful white formal double camellia called 'Alba Plena', and a red-and-white-petalled one called 'Variegata', arrived from Asia in 1792 and caused tremendous excitement throughout European horticultural circles.

The fascination of European nurserymen and horticulturalists was increased when, in 1820, the first Yunnan reticulata camellia arrived in Britain, having been named after the captain of the trading ship which transported it. It was the now famous 'Captain Rawes', and was followed by the plant-hunter Robert Fortune's namesake, *C. reticulata* 'Robert Fortune', which also goes by the name 'Pagoda', just to confuse us. These two beauties are still widely grown today, testifying to the fine quality of some of the older Asian varieties.

By this time the cultivation of camellias was widespread throughout Europe. Italy, Portugal, Spain, Belgium, France, Germany, Britain, and even distant Russia, all had thriving industries, and by 1830 a great boom had occurred with thousands of seedlings being raised by enthusiastic nurserymen. Alfred Chandler, nurseryman of Vauxhall, England, was typical of these keen hybridisers, and in 1831 he introduced his famous 'Elegans', which was the ancestor of a notable family to which belong today's favourites, 'Elegans Champagne', 'Elegans Supreme' and 'Hawaii', as well as many others.

The camellia was popularised by the French writer Alexandre Dumas, whose novel *La Dame aux Caméllias* was later adapted to opera and film. By 1892, only 100 years after its arrival in Europe, it had become the basis of a thriving cut-flower industry, centred on Nantes in France, which was able to supply Paris with 100,000 blooms of 'Alba Plena' for New Year buttonholes.

Towards the end of the nineteenth century camellias went out of favour, and the belief that they were hothouse plants could have contributed to this. In England, the end of the Victorian era saw the temporary eclipse of its fashionable floral symbol, the camellia.

Chapter 2

CAMELLIAS THIS CENTURY

'Margarethe Hertrich'

THE camellia's decline in popularity at the turn of the century was not so marked in France and Italy as it was in England, due in part to the persistence and initiative of Henri Guichard of Nantes and his daughters, the Guichard Soeurs. New varieties continued to be introduced, and at the Chelsea International Flower Show of 1912 these French growers exhibited camellias, demonstrating to the world of horticulture that these plants were not too tender, fragile and difficult to be grown in the open.

However, the great era of popularity and expansion which has marked the present century was given its impetus by the introduction of a new species from China called *C. saluenensis,* which was hardier than *C. japonica,* had a much longer flowering period and was self-grooming.

Between 1917 and 1919 George Forrest brought back the seed of the species from China, and private gardeners and nurserymen began crossing it with *C. japonica.*

The *williamsii* hybrids were produced by Mr J.C. Williams of Caerhays in Cornwall. The camellia called 'J.C. Williams' after Mr Williams, and bred by him, was given a First Class Certificate by the Royal Horticultural Society in 1942 and an award of Garden Merit in 1949, stating that 'it should be excellent for ordinary garden decoration'. A suitable long-flowering and hardy outdoor camellia had been achieved. Another famous cross of *C. saluenensis* with the *C. japonica* hybrid, 'Donckelarii', called 'Donation', was bred at this time by Colonel Stephenson Clarke at Borde Hill in Sussex. This camellia became one of the most

widely popular in the history of camellias and is still very much grown today. It should be noted here that only crosses between *C. saluenensis* and *C. japonica* and its cultivars are correctly named *williamsii* hybrids, although many successful crosses have been made between *C. saluenensis* and other species.

Plants of *C. japonica* are known to have arrived in Australia in 1831 and from there spread to New Zealand in the same year. In the United States the arrival of the first camellia has been estimated by historians to be 1797 or 1798. One American nurseryman listed 17 different cultivars or varieties in 1822. In New Zealand and Australia some of the best *williamsii* hybrids have been produced. Notable breeders include Les and Felix Jury in New Zealand and Professor E. G. Waterhouse in Australia with hybrids such as Les Jury's 'Anticipation', 'Debbie', 'Elsie Jury', 'Elegant Beauty' and 'Jury's Yellow', and Felix Jury's 'Waterlily' and 'Dreamboat'. From Professor Waterhouse

'Cinnamon Cindy' (above) and 'Scentsation' (above right) are two of the popular older scented varieties.

came 'Margaret Waterhouse', 'Lady Gowrie', and the most famous of all, named after him, 'E. G. Waterhouse'. A pioneering group of American hybridists contributed some interesting work in this field. Notable among these men were Dr Clifford Parks, W. E. Lammerts, J. A. Asper and David Feathers.

Because of the good cooperation and communication between nations, these enterprising men inspired others, and both amateur and commercial breeders began experimenting with different crosses so that new varieties continued to be developed, adding great diversity and interest to the developing genus.

In the last 30 years scientists have concentrated their efforts in hybridising on a search for fragrance. The best results have come from using the species *C. lutchuensis*

catalogues of scented camellias.

Extending the colour range has been another aim of hybridists this century. We now have a spectrum from white through all shades of pinks and reds to an almost black-red, but the discovery in China of the yellow-flowered species *C. chrysantha*, and several other associated yellow-flowered species, has inspired a strong hope that yellows and apricots can be added to the colour range, although up to now little success has been achieved.

The search for a pure blue-coloured flower, although not fully successful as yet, has led to the addition of several camellias with lavender and purple overtones in their red or pink shading. Some of these lavender-toned ones are 'Softly', 'Persuasion', 'Pink Dahlia', 'Neil Armstrong', 'In the Purple', and 'Purple Gown'. From New Zealand comes Mrs I. Berg's 'Blue Bird' and from America come 'Blueblood' and 'Blue Danube', while an older Portugese variety is 'Donna Herzilia de Freitas Magalhaes'.

There has been a growing interest in smaller, slower-growing varieties suitable for city courtyard gardens. Their smaller leaves and flowers, and their suitability for standardising, espaliering and container growing have extended the range of uses and are in keeping with the smaller scale of planting required in a city garden. Here, too, good work has been done in the Southern Hemisphere countries. Australia's Tom Savige has produced the popular hybrids with miniature flowers, 'Wirlinga Princess' and 'Wirlinga Gem', as well as 'Wirlinga Belle'. These dainty camellias, with differently formed pink flowers and spreading growth, make most attractive plants for smaller gardens or enclosed courtyards. In Australia, Edgar Sebire contributed new and attractive hybrids, namely 'Sprite', 'Snow Drop' and 'Alpen Glo'. This trio are all particularly free-flowering.

Neville Haydon in New Zealand pro-

as a parent, but promising results have also been obtained using the recently available species *C. yuhsienensis*. Other fragrant species which are being tried as parents in hybridising are *C. fraterna, C. tsaii, C. kissi, C. sasanqua* and *C. oleifera*. Some of the older fragrant varieties are 'Odoratissima', 'Kramer's Supreme', 'Christmas Daffodil', 'Cinnamon Cindy', 'Scented Gem', 'Sugar Dream', 'Scentuous' and 'Scentsation'. Many of the *C. sasanqua* varieties also have a pronounced fragrance.

In the search for fragrance, American, Australian and New Zealand hybridists have all contributed to progress made. Dr Clifford Parks, Dr Ackerman, David Feathers, the late Robert Cutler and the late Kenneth Hallstone, were all involved in study and breeding programmes in America. In Australia, Ray Garnett and Edgar Sebire have both been successful with intensive hybridising programmes. Jim Findlay in New Zealand has produced valuable scented hybrids including 'Katie Lee', 'High Fragrance', 'Superscent' and 'Scentuous'. His countrymen, Os Blumhardt with 'Sugar Dream', John Lesnie with 'Quintessence' and Trevor Lennard with 'Gay Sue' have all made valuable contributions to the

'Wirlinga Princess'

The truly dwarf-growing 'Baby Bear.'

duced the deservedly popular 'Baby Bear' and 'Baby Willow', which have a true dwarf growing habit and delectable tiny flowers. John Lesnie's 'Quintessence' is valuable both for its fragrance and as a container plant.

One of the most valuable cultivars to be raised in New Zealand in recent years, in the opinion of many, has been Bettie Durrant's 'Nicky Crisp'. This new camellia is a slow-growing *C. pitardii* seedling which has pale pink, semi-double flowers of great delicacy. It grows to a height of about 1 m and I have used it successfully in my design work as a charming internal hedge to surround a planting of smaller rhododendrons. Other useful slow-growing seedlings of *C. pitardii* produced by Mrs Durrant have been 'Snippet', 'Prudence', 'Persuasion' and 'Contemplation'.

Although the most widespread hybridising during this century has been done in the United States, Australia and New Zealand, one outstanding British hybridiser in Cornwall, that southern county which is so famous for its magnificent camellias and rhododendrons, was the late Gillian Carlyon. Living at Tregrehan Garden she produced, from 1960, some fine crosses of *C. saluenensis* and *C. japonica*. Notably successful have been 'China Clay', 'Gwavas', 'E. T. A. Carlyon', 'Edward Carlyon', 'William Carlyon' and 'Tregrehan'. The UK has also produced several good contributions from the Trehane family, namely 'Joan Trehane' and 'Jennifer Trehane'.

Chapter 3

THE NEEDS OF CAMELLIAS

'Tui Song'

GOOD gardeners try to emulate in the garden environment the conditions under which plants thrive best in their natural habitat. Camellias in the wild are found growing on slopes, in thickets, in forests or in more open woodland areas. There they have light shade from overhead trees, the fairly shallow layer of soil is acid, containing plenty of organic material in the leaf fall from the overhanging trees, and, above all, the drainage is good.

Shade

Woodland plantings in larger gardens approximate the natural habitat of the camellia, and are an ideal way to grow them where space is available. Tall deciduous trees are best for creating high overhead shade. They also provide protection from hot winds, the scorching rays of the summer sun and from extreme cold in winter. In the woodland garden do not plant shade trees too closely or choose varieties with very invasive root systems. This will cause root competition and thus deprive the camellias of adequate water and food. Lack of adequate light will prevent bud set, so make sure that there is sufficient filtered light for the sheltered camellia's full health. (See Chapter 9 on landscaping for suggestions for suitable shade trees.)

Soil requirements

The ideal soil for camellias is a free-draining, slightly acid one which is rich in humus. These plants, however, are extremely adaptable and will tolerate a wide range of conditions. In defining the term 'slightly acid', it is necessary to describe the pH scale which shows the degree of soil acidity and alkalinity. This scale has a range of 0 to 14, with 7 indicating a neutral state. Numbers below 7 show acidity and numbers above 7 indicate alkalinity. The term to most accurately describe the camellia's preferences is 'acid tolerant', as they thrive best in soil measuring pH 6 to 6.5, and while a little calcium is desirable, they will not stand very much of it. It is easy to assess the approximate soil acidity of a new property by noting what is growing next door. If there are healthy looking rhododendrons, daphnes, camellias, ericas and other acid-tolerant plants you may be fairly certain that your soil approximates the required acidity for their health.

There are many gardeners all over the world growing camellias well on soil that has never been tested. Some gardeners, however, desire a more accurate measurement, especially when making a new garden with a substantial outlay of capital on plant material. Expert commercial soil-testing services are available at a price, but there are also several useful 'do-it-yourself' kits on the market which are readily available to the home gardener from nurseries and garden centres. If, on testing, the pH reading is found to be too alkaline, i.e., above pH 6.5, perhaps because of previously applied lime and not entirely due to an alkaline subsoil or underlying rock layer, the condition can be corrected by the addition to the soil of ferrous sulphate. This has been found to be cheaper than pure sulphur, it acts more quickly and reliably and can be readily obtained from local garden centres.

A useful guide to quantities needed to reduce alkalinity of the soil can be found in David Leach's book *Rhododendrons of the World:* 'It requires 18 pounds of crude ferrous sulphate per 100 square feet to reduce the pH from 8 to 6. It requires 16.5 pounds to reduce the pH from 7.5 to 6 and it requires 9.4 pounds to reduce the pH from 7 to 6. The pH needs to be tested again a few days after the chemical has been sprinkled over the soil and watered in.' It is wise to use only acidic fertilisers such as ammonium phosphate, superphosphate, sulphate of ammonia or sulphate of potash. If there is possible runoff from nearby building foundations or from a well-limed vegetable garden, it is advisable to spread around a little ferrous sulphate each year.

Although camellias are 'acid tolerant', they can react adversely to a very strongly acid soil. This is because important minerals in the soil particles, mainly calcium, manganese and some iron, may become unavailable to the plant in strongly acid conditions. Then adverse symptoms may appear, such as malformed, small foliage with leaves becoming margined and blotched dull yellow, growth shoots dying and falling off or slow, thin, weak growth dying back from the terminals. Dolomite is better to use to correct the calcium requirements for the camellia, as they dislike free lime.

Humus

The regular addition of humus or organic material will make the soil more acid, i.e., lower the pH of the soil in a natural way. Humus also improves soil texture and water retention, and during decomposition provides extra plant food for the camellia. The addition to the topsoil of rotted sawdust, peat moss, garden compost, leaf mould and very old cow manure, with a ratio of three parts of topsoil to one of humus, is a good guide to enrich the soil. Porosity, i.e., good aeration of the soil, is a prime requirement for a healthy camellia plant. To improve the soil texture of a very heavy, puggy soil, a 50/50 mixture of soil and humus, to which some sharp river sand has been added, is advisable.

Camellias enjoy a woodland-type garden where the soil has regular applications of humus, or organic material.

Drainage

Drainage and rainfall are important factors bearing on good soil condition, the necessity for fertiliser and the amount required, and any improvement to soil texture that may be called for. Nutrients will be leached from the soil more quickly in high rainfall areas, so that camellias will need relatively more food here than in drier areas. If, in addition, in these high rainfall areas the soil is porous, sandy and light, the addition of organic material to help retain moisture and nutrients will be doubly important.

A camellia's growth will be seriously limited by a heavy, puggy, waterlogged soil.

Even after three to five days of waterlogging, oxygen is unable to enter the pores of the soil and any camellia growing in these conditions will become unhealthy as a consequence. Waterlogged conditions also create an ideal environment for the development of fungal diseases, including *Phytophthora,* which can prove fatal and at best can badly weaken the plant. The drainage should be thoroughly investigated, and field drains laid if necessary, prior to planting a new area with camellias. If the soil is extremely badly drained it will have a characteristic 'rotten egg' smell and a rather nasty pale grey colour. As mentioned earlier, heavy clay soils can be completely changed over a long period by the frequent addition of plenty of organic material, sharp sand or grit.

The creation of raised beds is another

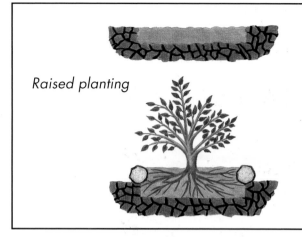

Raised planting

Preparation of site for raised bed. Excavate the site of the bed to a depth of 50 cm and fill with new free-draining soil of a suitable acidity.

The completed raised bed. Raise the bed above the surrounding level by at least 50 cm, using logs, tanalised timber or old railway sleepers, and fill with new free-draining soil of a suitable acidity.

solution to poor drainage. They can also provide the correct environment on an area of alkaline soil, as the new soil in the raised beds can be adjusted to suit the camellias' requirements, i.e., can be made slightly acid. Any suitable edging, such as tanalised timber or old railway sleepers can be used to raise the bed the required half a metre above the surrounding soil. (See diagram on 'Raised planting' above.)

Soil maps can be very helpful in understanding local water tables, drainage patterns and soil types which can cause waterlogging problems. These may be able to be obtained through a government department dealing with land information or a local council.

Chapter 4

CHOOSING AND
PLANTING CAMELLIAS

'Fairy Blush'

MAKE your purchase from a specialist camellia nursery or from a well-established garden centre or general nursery where the staff know their plants. They can then give you sound advice on the choice of variety to suit your climatic area, your garden's soil type, and your personal requirements as to flowering time, colour, bush size and shape. An experienced nurseryperson is invaluable, as he or she should know which camellias to suggest to achieve the garden style you wish, be it a woodland garden, a cottage garden, a modern court-yard garden, or a formal garden containing espaliered and standard camellias.

Guidelines for buying wisely
1. Avoid plants showing any sign of disease.
2. Choose a plant with leaves of a rich green colour and good texture and at least 45–60 cm in height.
3. Don't buy any plant that has grown tall and spindly in its container. It has probably been there too long.
4. When buying a plant in flower, look for good colour, texture and substance in the flowers.
5. Beware of plants that have outgrown their containers, those that have been too long in the nursery or those that are in the bargain bin!
6. The position available for planting in your garden will affect your choice. If the camellia is to be planted where it is ex-posed to frosts and bad weather, it is better to choose one of the deeper-coloured vari-eties as whites and paler colours will be more obviously badly damaged by adverse

Shelter from the wind is desirable for all camellias.

weather conditions. **NB Never plant a camellia where it catches the morning sun.**

7. Increase your own knowledge of camellias. Attend local camellia shows and visit the botanical gardens in your area. Join a camellia society. Most societies compile lists of varieties that thrive in their local areas, and also of varieties that have been well tried.

Planting: where and how

1. Choose a planting site to suit your particular camellia variety. Japonicas prefer a semi-shaded position, while the sasanquas and reticulatas will tolerate more exposure to sun and light.
2. Wind shelter is desirable for all varieties.
3. The site should be free from the invasive roots of nearby trees.
4. Some summer sun is desirable as too much shade will prevent good bud set.
5. A regular and adequate supply of water is necessary, as is good drainage.
6. Spacing is most important. Gardeners

often tend to plant much too closely, forgetting that a cluttered, overcrowded look is not desirable. If this does happen it must be remembered that, with care, even mature camellias may be shifted successfully. But it is better to plan at the outset rather than be faced with expensive moving operations in a few years' time.

It is unwise to plant japonicas closer than 2 m apart, even though judicial pruning keeps them in bounds to some extent. The larger growing reticulatas, which are taller and more spreading than the japonicas, need 3 m between them, and specimen trees planted in isolation in the lawn need at least 5 m to develop their natural form.

Planting method

1. An overnight soak in a large bucket or tub is advisable for the new camellia before the planned planting. The camellia should then be left long enough for surplus water to drain away before planting.
2. Having established that drainage and basic soil type are suitable, prepare a hole at least four times as wide and three times as deep as the size of the planter bag in which the camellia arrived from the nursery. It is better to prepare a little more, rather than a little less, of the surrounding soil. This will ensure that the camellia will have plenty of friable soil for its new feeding roots and thus have a really good start.
3. Mix thoroughly into the bottom of the hole a mixture of one-third compost, two-thirds soil and a handful of slow-release fertiliser.
4. After firming down a little, adjust the level of soil so that the plant will be slightly above the surrounding ground level, thus allowing for the settling of the earth after watering and also assuring that the drainage will be good.
5. A stake should be put in position now if the plant needs support. This will avoid driving the stake through the root ball after

Correct planting

Roots are spread and plenty of friable planting mix has been added around the roots. Mulch has been added above soil level.

planting.

6. Try the plant in the hole to check that the back fill is at the correct level before removing it from the container.

7. Position the plant in the hole, still in its container, and when you are thoroughly satisfied that level and placement are correct, slit the plastic bag down the side with a sharp knife and slide it out from underneath the plant.

8. Unhealthy looking brown roots should be carefully trimmed off with a sharp sterilized knife before planting. Healthy roots should be white and these can be carefully teased out of the root ball and spread out in the hole.

9. The hole can now be filled with the same soil and compost mix and firmed down gently to exclude air pockets.

10. After watering in your camellia thoroughly, the job will be complete.

(See diagram above on 'Correct Planting'.)

Transplanting

The fact that camellias can be transplanted when they are fully grown is one of their great advantages. Landscapers can provide clients with an instantly mature-looking garden, and the private gardener, when taking over an old-established property, can move the mature camellias and rearrange the garden to taste, so saving on expensive replacements. Mechanical diggers can be used to move very large trees, but it is more usually the task of moving garden shrubs between 1 and 2 m high in order to change the garden layout which concerns the home gardener. Optimum transplanting time for camellias is during their dormant period, i.e., from late autumn through winter to early spring. In warmer climates where spring growth takes place earlier, it is best to complete transplanting tasks by the end of winter, well before the start of new growth.

Wrenching

It is wise to wrench an established camellia well before the planned move, and because the root system will be a good deal smaller after wrenching, heavy pruning will be necessary before transplanting takes place.

Wrenching consists of making a vertical cut in the soil all round the tree with a sharp spade, just inside the spread of the outer leaves. This can be done in autumn if the move is planned for winter or spring. If the camellia is very large, has only one or two thick roots, and lacks a cohesive root ball, cut round the roots in spring to encourage fibrous roots to form, ready for lifting in the autumn. A working trench can be dug just outside the cut to the depth of the spade, to allow for leverage to enable the root ball to be separated and undercut all around the plant.

A Cornish shovel with a long handle can be used to tilt the root ball up on one side, so that a piece of scrim or sacking or a strong polythene sheet, can be slipped underneath. After sticking a fork into the root ball to guide and steady it, wriggle the whole thing on to the sacking and then drag this, with the camellia on it, to the new site.

Plant the camellia at the same depth as

before and position it as it faced before, otherwise leaves that have been shaded from the sun in the past may now be badly scorched. Pack damp peat around the plant and fill in the hole. Keep the bush moist during transplanting and water regularly for the first few weeks after the move.

Mulching

The natural leaf fall accumulating on the forest floor creates a perfect mulch for camellias growing in their wild state. This protective blanket of leaves and organic material provides many advantages for the developing camellia plant — controlling weeds, retaining moisture, providing humus and some nutrients, and modifying soil surface temperature. It also prevents erosion by rain and helps reduce the depth to which frost will affect the ground.

Wise gardeners, seeking always to recreate the natural conditions in which their chosen plants thrive, can do no better for camellias than to provide them with a good mulch as soon as they are planted, and then keep adding material so that a good continuous protection is provided. Mulching is particularly important in the first few years of the camellia's life, when its root system is becoming established. Suitable mulching materials include pea straw, pine needles, rotted oak leaves, chopped bracken, untreated bark chips, spent hops or well-rotted sawdust. (NB If fresh sawdust is used, a sprinkling of sulphate of ammonia must be added to the mulch to replace the nitrogen extracted from the soil during the rotting-down process.)

Live mulches are another option open to the camellia enthusiast. These can be any groundcover of the surface-rooting variety. There are fierce differences of opinion among the experts on this topic. Some say that many groundcovers become too invasive and compete with the camellias for available nourishment, while others heartily endorse their use.

A mulch of stones acts as an efficient weed-suppressor, keeps roots cool and conserves moisture in the soil. It has the extra advantage of adding an attractive grey textured pattern to the garden scene.

Beware of unsuitable mulches which can actually damage the health of the tree. For example, fresh grass clippings on their own become soggy and can exclude air from the root system. Note too that the leaf mould from certain trees is not beneficial, for it produces an alkaline reaction when it breaks down, e.g., elm, ash, sycamore, horse chestnut and lime trees.

Applying the mulch

Be generous when applying mulch as it settles and rots down very quickly in many cases. Pile it over the root run to a depth of at least 8 cm, but be sure it does not lie against the trunk of the camellia bush, where it may cause bark rot. Christchurch's Edgar Stead, who was an expert New Zealand grower and hybridiser of acid-loving evergreens, wrote in an article in the Royal Horticultural Society of Britain's *Rhododendron Yearbook* (1947) that mulching was one of the best protections from the fierce sun of the Southern Hemisphere and the hot, drying winds that prevail there during the summer months.

Mulch may usefully be applied twice yearly: in spring, when the soil is warming up but still moist after the winter, and again in the autumn, when the soil is still warm from the summer, but dampened by the moister weather of autumn.

Watering

While perfect drainage is essential for the camellia plant, the soil should be kept constantly moist, but not saturated. Adequate water is particularly important to a newly planted camellia to ensure that it has the best possible chance of establishing well in

A protective blanket of leaves and organic material will help to recreate the camellia's natural habitat.

its new position. Water well in dry weather and make sure that the soil under the mulch remains damp. Plants growing under the eaves of houses need particular attention when watering, as do those growing near or under big, moisture-robbing trees and hedges, or at the feet of dry walls. It is worth considering a soak hose or an automatic watering system for these vulnerable positions. Conditions that are too dry not only harm young plants, but are a major cause in mature plants of a failure to set flower buds for the new season, and also of bud drop in winter.

Rainwater is the best for watering, but failing this, the effects of hard tap water can be nullified by the use of sulphate of ammonia as a fertiliser.

Variables such as soil type and climate make it almost impossible to lay down hard and fast rules for definite watering programmes, and they will alter from district to district and from country to country. Local knowledge and advice from experts must be used, as well as common sense and discretion, but generally the goal to aim for is a soil kept moist but not saturated at all times.

Camellias should not be watered from above on a very sunny day, as this can cause leaf scorch, which produces ugly brown blotches all over the foliage. After a hot summer's day, a good sprinkling of the foliage in the cool of the evening not only discourages insect pests but washes off dust and raises the humidity. As a general rule, one or two good soakings a week are more beneficial than more frequent light, patchy applications.

Feeding

Camellias give a more generous return for good feeding than most other plants and a sumptuous array of flowers and glistening, healthy foliage are a great dividend for sound feeding methods.

One important point to note at the outset of this discussion on food, is that it is better to under-feed than over-feed. More young plants die from over-feeding than from neglect, so it is always better to err on the side of frugality with fertiliser application. It is very simple to apply a little more feeding at some later date if it is thought that an application has been too light, but it is very difficult to restore a camellia plant which has been badly scorched by an overdose of chemical fertiliser.

The other important point to note is that it is never wise to apply any sort of fertiliser to dry ground. Always water before applying fertiliser and again afterwards, to water the food into the mulch.

Advice here must again be of a general nature only, as soils and growing conditions vary greatly from area to area. Local advice from camellia societies and neighbouring gardeners is advisable for the fine tuning that will be necessary.

The aim in fertilising is to produce and maintain the slightly acid soil which has been described earlier. This can be best achieved by the addition of natural organic material. This takes a little more trouble but provides a healthy environment for camellias and associated garden plants and is less expensive than chemical fertilising. Unlike chemical fertilisers, which stimulate for short periods only and then fade out, nutrients from organic sources become available more slowly and continuously. The best

Feed your camellias generously and they will respond in kind, as 'Anticipation' has done here.

asset for the organic gardener is a well-made home-compost bin and there are many manuals to advise on correct methods of composting.

Supplement your homemade compost with other natural manures such as blood and bone, medium ground hoof and horn meal, well-rotted stable or cow manure, or well-composted sheep or fowl manure. If manure is first composted you will reduce greatly its viable weed seed content.

When using chemical fertiliser, choose one of the proprietary brands on the market today that have been specially formulated for acid-tolerant plants like camellias, rhododendrons, daphnes, etc.

Chapter 5

PESTS AND DISEASES

'Gay Baby'

IN general, camellias are not much subject to attack by insect pests, although in some areas of the world where conditions are warm and humid they can be more of a problem than in countries which have cooler temperatures and sharper winters.

Pest prevention

1. Good soil health ensures a healthy plant which will be resistant to insect and disease attack, so concentrate first on building a healthy soil before planting camellias.

2. 'Sanitary' pruning to remove dead wood, prevent overcrowding in the bush centre, and allow for better air circulation, will greatly reduce the risks of insect infestation and make it easier to deal with, should it occur.

3. A high-pressure lawn sprinkler gives a very good clean-up of thrips and scale insects which both inhabit the underside of leaves. The pests are literally washed away if the sprinkler is left running for an hour or so under a properly pruned bush.

Chewing insects

Types of chewing insects vary from country to country. They include snails, grasshoppers and katydids and some night-feeding varieties.

Bronze beetle This beetle is a night flier. Young camellia leaves will have been found scalloped by an invisible assailant early in the morning. If you investigate at night with a torch you will find that a team of small flying beetles is working its way from one camellia bush to the next.

Remedies Use an insecticide containing acephate, nicotine sulphate or carbaryl. The pyrethroids are also reasonably effective. Fortunately most beetle flights take place in spring and early summer so that the problem largely disappears once the leaves are fully expanded and hardened.

Grass grub beetle This beetle also flies at night. The adult is shiny and golden-brown with a hard body. It is active for about 4 weeks during the last month of spring and the first month of summer. Young camellia leaves, and those of many other plants, are included in the diet of this pest.

Remedies Starlings are natural predators. Pesticide granules, spread on the lawn, will reduce the numbers of grubs which will later hatch into beetles and chew leaves. These beetles hate water, so a strong spray of water left playing on camellias at night during the weeks when the beetles are active is a good preventative.

Leaf roller caterpillar These caterpillars belong to three different species of moth which build tents for themselves by making webs where two leaves overlap, or by rolling up soft new leaves. They can do a great deal of damage in a short time to young growth as they are hungry leaf-eaters.

Remedies Because leaf and web covering often prevents insecticides from reaching the caterpillars, a good squeeze between the fingers is a most effective control. The pyrethroids are most effective against them, as are acephate solutions and many other sprays if properly applied. Consult your local garden centre.

Bag or case moth The larvae of this moth are responsible for damage to camellias. The bag-like structures which they make for protection are an easy means of identification.

Remedy The most effective method of control is to remove and destroy the cases by hand.

Sucking insects

Aphids These pests are active when the plant is growing strongly. They are usually found on new buds or shoots but may feed on the underside of older leaves. They are very small. Symptoms include twisted and wilted shoots and leaves and buds which produce distorted flowers or may fail to open. Honeydew from aphids and associated scale often results in an unsightly black coating on the leaves, called 'sooty mould'.

Remedies Hosing is effective, but needs to be repeated as each new batch of aphids appears. Numbers may be reduced by their natural predators — wasps, ladybirds, lacewings and hoverflies. Soapy water will kill insects on contact but will not be effective for more than a day. A systemic insecticide is effective. Ask your nurseryman for advice.

Scale insects Some scale insects are hard-shelled and some soft. They vary in shape, size and colour, each resembling a miniature limpet about 1–2 mm long, and may cause considerable damage to camellias. They are found on the under-surface of older leaves, thrive under very dry conditions and may be white, brownish or even dark brown. Affected camellia leaves become yellow if sun-scorched and are unable to supply sugars to the plant any longer.

Remedies Ladybirds and parasitic wasps are natural enemies of the scale insect. In severe cases it may be necessary to prune and burn badly affected branches. Chemical control can be best achieved by the use of an all-purpose spraying oil. Note that an insecticide such as nicotine sulphate, which can be added to the spraying oil, is a contact poison of relatively low toxicity to human beings but is highly toxic to bees and other beneficial insects. You should therefore use it with discretion, weighing up the possible advantages against the obvious disadvantages.

Thrips Camellia leaves affected by the rasping feeding action of thrips present a typical silver appearance.

Remedies An effective insecticide is nicotine sulphate. Thrips often migrate to camellias from a nearby host so it is wise to look for shrubs in the vicinity which show signs of thrip activity and treat these also.

Mites These pests strictly speaking are not insects but arachnids. There are at least three different varieties of mites which have been found on camellias. They are too small to be seen with the naked eye but show up well under a strong magnifying glass. Damage to camellia leaves is obvious, for mites cause a reddish dusty, grey dusty or silvery appearance on the underside of the leaf. With a severe infestation, the upper side of the leaf can develop a similar appearance.

Remedies These pests are hard to eradicate with standard insecticides because they are not true insects. Ask your nurseryman for a good miticide. Some miticides are available which kill active mites and others kill the eggs. It is necessary to spray the eggs at least 3 times at 7–10 day intervals, preferably with a change of miticide on the second and third spray. This should get rid of an established infestation.

Mites are much more of a problem when the air is hot and dry and the plant is short of water, so a regular misting of the leaves during dry periods, or a spray with soapy water, makes a much less attractive habitat. Soap acts as a weak miticide. Oil sprays will slow down mites greatly if applied before an infestation has become severe. If a camellia is in a position that is too hot and dry and mite infestation is severe, it is worth considering transplanting to a cooler, moister place in the garden.

Mealy bug This insect is approximately 3 mm long, with clearly defined segments and legs. Whitish in colour and covered with a white waxy powder, it is sometimes first detected by a cottony material which it secretes. It sucks sap and secretes honeydew like the scale insect but is mobile. It typically hides in crevices for much of the time. Camellias in tubs may be more subject to infestation by mealy bug. Control is the same as for scale insects.

Tunnelling pests

Leaf miners The larvae of several types of insects make the leaves of camellias unsightly. The larvae tunnel inside the leaves, leaving unattractive 'mines'. Sometimes these look like blisters and sometimes like thin, wavering lines.

Remedies Use a systemic insecticide to eliminate these pests.

Stem borers The grubs of several different insects burrowing in the wood may damage camellias. Small piles of sawdust, which they push out of the holes they make, indicate the presence of these pests.

Remedies include using a small bottle of methylated spirits fitted with a finely pointed top. A quick spurt of meths into fresh borer holes deals with the pest on the spot. Alternatively you can syringe the holes with soapy water or weak insecticide. As a last resort, the whole affected branch or branches may have to be pruned off and burned.

General nuisances

Ants and earwigs Although these do little harm to camellias, they are a nuisance in the garden. They can be helpful in locating a hitherto undetected infestation of scale or aphids, being often attracted to the honeydew. However, ladybirds which are the natural predators of scale and aphids, avoid the habitat of ants, so natural control methods are inhibited by the presence of ants.

Remedies A good control for ants and earwigs is to sprinkle pesticide granules at the base of the camellia bush. A grease band applied low down on the trunk of the camellia will keep these insects away from the flowers.

Slugs and snails These molluscs usually

pose only minor problems for the camellia grower but if there is a likelihood of them attacking a prize bud just as it is opening before the annual camellia show it is useful to know of a preventative.

Remedy Snail baits based on metaldehyde sprinkled around the base of the trunks of camellia bushes are effective, but can be a danger to domestic pets. Prevention is much better than cure where slug and snail control is concerned.

Cicadas Female cicadas can damage camellias by cutting into their branches to lay their eggs, usually leaving a herringbone pattern.

Remedy Prune branches soon after they have been attacked to remove the eggs. There is no chemical control.

Diseases

Camellias suffer from relatively few diseases, unless they are mishandled, but in each country there are pathogens which thrive in particular local environments and manifest themselves when conditions are favourable. By far the best defence against camellia disease is to maintain optimum soil conditions and provide adequate water to ensure full health and vigour in the plant. Healthy plants will be better able to resist disease.

Phytophthora cinnamomi (root rot) This is the most serious camellia disease as it attacks all ages, from tiny nursery plants to mature trees. *Phytophthora* flourishes in heavy, waterlogged soil with poor drainage so the best preventative measure is to plant in free-draining soil or a raised bed.

Symptoms to watch for include: drooping of the leaves, yellowing of the leaves, wilting and leaf drop, branches and stem tips withering and dying until eventually the plant itself is dead; if a small piece of bark is cut away from the base of the trunk, a dark stain may be seen in the wood; many of the roots will have brown/black lesions

on them and if the dead-looking lengths are rolled between the fingers, the cortex will slough off, leaving behind the thin, woody core.

Remedies The disease is difficult to control, so prevention is the goal. Always ensure perfect drainage. A suggested treatment which may save an infected camellia is to lift the plant and wash off any soil from the roots. Trim off any dead, brown-looking root material, cutting back to healthy white root fibre, then soak the roots in a good reliable fungicide for at least 20 minutes. The plant should be replanted in a different, well-drained part of the garden.

This procedure is impracticable for large plants growing in the open ground, but may help to save smaller or container-grown plants. Large plants can be treated with an application of a fungicide to the soil around their roots.

Glomerella cingulata (collar rot or dieback) Sasanqua camellias are more susceptible to this disease than other cultivars. Twigs and short branches first die back, and leaves rapidly turn brown and sometimes fall off. This is followed by a stage which is typified by the dying of tissue and the growth of a canker. If cankers on the trunk or branches continue to enlarge undetected, water and nutrients will be prevented from reaching the branch tops, resulting in yellowing of leaves, dieback of twigs and branches and loss of foliage.

Remedies Cut off all infected wood, making sure that the end of the cut branch is creamy-white. You must cut further back if there is still a brown stain evident in the stem, as this indicates that the infection is still present. Sometimes a sunken part of a branch indicates the presence of an infected section of the stem. It is possible to save a large branch by carving out the infected piece, but this is not always successful. Finally, spray the plant with a suitable fungicide to clean up any minor infected sites you may have missed. Fungicides containing benomyl, captan, or the copper sprays, are suitable to use.

If the problem tends to recur on plants in your garden, it would be a wise precaution to spray all camellia plants with one of the above fungicides immediately after you have pruned them, as the infection typically enters through pruning cuts and scars.

Exobasidium spp. (leaf gall) Leaf galls are wind-borne fungi affecting leaves when they are very young and they are common in many countries. In affected camellias whole or part of the young leaf, shoot tip or flower bud becomes white and puffy, rather like a cauliflower heart. These galls are very prominent and ugly and always at the ends of branches. Spraying is ineffective. Remove the diseased leaves and burn them, or tie up in a plastic bag and dispose of them in your rubbish collection.

Leaf spot These could be caused by *Pestalotiopsis,* a fungus that enters parts of the leaf which have been damaged by sunburn or in some other way, or *Monochaeta camelliae*, a more common fungus. Here the affected part may have black pin-head-sized bodies of a fungus showing against the unhealthy, silvery-grey collar of the infected leaf. Prevention relies on sound cultural practices to prevent leaf damage. It is important to remove and burn diseased wood and leaves. Careful treatment of your camellia bushes ensures that you will reduce the possibility of stress through injury which makes it easier for fungi to enter plants.

Ciborinia camelliae (petal or flower blight) This disease was previously known as *Sclerotinia camelliae* and affects only plants of the camellia family. Australia, New Zealand and Europe have been free of it until recently, but it has been prevalent in China, Japan and USA. Flowers only are affected. Small brown spots appear, and the disease spreads until the whole flower becomes brown. Veins on the petals darken so that the flower shows a netted affect which distinguishes flower blight from climatic injury. Also, if the completely infected flower is turned over and the bracts are taken off, a circle of greyish-white, cottony growth can often be seen around the base. Eventually the whole flower, ugly and brown, drops off. If there are favourable warm and moist conditions, the browning process can be as rapid as three days.

The spores of the disease are transmitted when hard, black, resting structures called 'sclerotia' are formed around the base of a fallen blossom. This takes about a month. These sclerotia remain in the leaf litter at the base of the bush which bore the diseased blooms until the following spring.

When the new crop of blossoms are beginning to open, the sclerotia germinate, forming large fungal structures shaped like trumpets which can easily be seen under infected trees. Some have stalks and others do not. These rather gruesome-looking structures produce the spores which re-infect the fresh blossoms for the new season. These spores are puffed out and carried by the wind, and once they touch a bloom, the germinating spores re-infect the blossom and the cycle is repeated.

Prevention No fully effective fungicide has yet been found which will eliminate the disease, so the best attack so far seems to be good garden hygiene to prevent or delay its spread and impact. The daily gathering up and burning of all fallen petals prevents the spread of the disease to some extent. Ensure good light and air circulation around the base of your camellia bushes by keeping the ground at their feet free of vegetation. Examine the bushes of neighbours to make sure they are not the source of infection. Explain how the disease spreads and encourage garden hygiene. Some American growers concentrate on container culture in an effort to avoid disease. Another preventative measure suggested is to wash off soil from the roots of new plants from the nursery and carefully and safely dispose of this growing medium.

Other problems

Chlorosis The presence of chlorosis is indicated by yellowing of the leaves. Evergreens shed their leaves systematically, as do deciduous trees. The difference is that the evergreens do not shed them all at once but drop some leaves each year. Before shedding, these old leaves normally become yellow, pale green, brown, or even red in mosaic patterns. The normal lifespan of a camellia leaf is about three years so there are usually some old discoloured leaves to be seen on the bush. This process is perfectly normal, but it is surprising how much alarm the sight of a few yellowing leaves can cause the uninitiated gardener. Other causes of chlorosis could be

(a) Yellowing of the leaves associated with the loss of chlorophyll. This may follow an excessively wet period when the plant roots have been damaged or are unable to function properly because of waterlogging of the soil. Conversely, it may be caused by a long spell of dry weather, when the plant's root system suffers from deprivation of moisture and is unable to support the bush adequately. If these situations are allowed to continue, the plant will eventually die, so improved drainage or a more regular water supply are needed.

(b) An excess of alkalinity in the soil is another possible cause of chlorosis. A high pH prevents necessary nutrients, such as iron, from being available in a soluble form. If drainage is good the condition can be remedied by an application of iron chelate or sulphate of iron. Compost, home grown if possible, or the use of one of the prepared 'acid' fertilisers would also be helpful.

(c) Green veins, surrounded by yellowing of the leaves, is typical of a lack of either iron or magnesium. Any of the following applications should bring a quick return to healthy colour and renewed vigour. (NB Fertilisers are best applied in small quantities.)

Apply either sulphate of iron, magnesium sulphate (Epsom salts), or a general fertiliser containing trace elements. Alternatively, spray with blood and bone containing trace elements, or apply a small quantity of dolomite lime.

Corky leaf (leaf scab) This problem, sometimes called 'oedema', shows up on the underside of leaves as brown blisters which burst and harden into scabs. It is thought to be a physiological reaction to conditions in the atmosphere or the soil. One theory is that overwatering may cause plants to absorb more water through the roots than the leaves can transpire. It can occur in camellias which have a small amount of foliage and a large root system, perhaps after drastic pruning. Well-drained soil and moderate, balanced feeding may prevent its occurrence. It does not seem to have any ill effects on the plant but looks unsightly and can be worrying. Don't remove affected leaves as this only aggravates the problem.

Algae In shady but humid positions, especially under pine trees, camellia leaves may become encrusted with a fine grey-green deposit which is difficult to remove. This does not seem to affect the health of the plant, although it is unattractive and severe deposits prevent light reaching the leaf. Pruning to open up the plant and make more light available would be a good preventative measure.

The incidence of algae may be reduced by spraying with copper oxychloride, and soapy water, sprayed alternately with copper mixtures, is also found to be effective.

Lichens Lichen is not a parasite and has no detrimental affect on the growth of the camellia, but it does look unsightly and could harbour insect pests. Some growers believe that it only appears on less healthy plants and so it is wise to give more attention to the general health of the bush. A buildup of lichens is an indication of slow growth, not a cause. Copper sprays (copper oxides, Bordeaux mixture) can give it a knockback, and lime sulphur is effective on other plants, so would be worth trying on camellias.

Viruses White blotches appearing on red flowers, and white variegations on foliage of camellias indicate the presence of viruses. White flowers do not show any symptoms. Informed opinion has not yet found any unassailable evidence that viruses are spread by insects, although British opinion claims that sap-sucking insects and scale are guilty of transferring viruses to camellias. There is no firm evidence either that dirty secateurs are responsible, but it is a good sanitary precaution to ensure that all tools are kept thoroughly clean and sharp. Viruses are definitely introduced by grafting on to infected stock, and by using scions from infected stock as cuttings.

Viruses are viewed differently by different countries. In France every precaution possible is taken to eliminate viruses, and infected plants are destroyed. In the United States, however, variegated plants are so much admired that some growers are quick to introduce a virus to plain-coloured culti-vars by grafting on to infected stock. In general, viral infection does no harm to the plant except that yellow leaves may be affected by sun-scald more easily.

There is no treatment or cure for viral infection. The best prevention is to try always to obtain virus-free stock from the nursery, but as we have discovered, when there is deliberate grafting on to infected stock, this is very difficult to ensure.

Notes on the safe use of poisonous sprays

1. All sprays should be applied strictly according to the package instructions and repeated as prescribed.
2. Spring and summer are the usual periods to spray.
3. When applying, choose a calm, overcast, or cool day so that the spraying material will begin its effective action before drying off. This applies particularly to white-oil sprays, which can cause sudden defoliation if applied too strongly on a hot day.
4. As most of the newer sprays are very effective but highly potent and toxic, protective clothing should be worn at all times to prevent contact with the skin.
5. Always read instructions very carefully and follow them to the letter.
6. Regular spraying to keep plants healthy and clean is much better than applying heavy doses after allowing plants to become badly infected.

Chapter 6

PRUNING

'Spencer's Pink'

AN unpruned camellia presents an unsightly, shapeless, neglected appearance and becomes an ideal sanctuary for fungal diseases and insect pests, because of overcrowding of branches and leaves and lack of air circulation in the centre of the bush. Maintaining good health in your camellia collection is therefore a major reason for regular pruning.

Also, with judicious pruning, the camellia flowers, which are particularly vulnerable to damage from weather, are better protected. It has been found that the flowers on a well-pruned camellia will tolerate considerable wind and rain and still retain the quality of their blooms. A more open bush, which is achieved by good pruning practice, allows blooms to move better in wind and wet weather without the abrasive action of over-crowded leaves and branches.

It is easier to train a camellia to the desired shape and size to suit your garden style and ideas, if pruning is started when the plant is young, and repeated regularly. With regular pruning you are in control, and to some extent can determine the ultimate mass to suit the camellia's position in the garden and relationship to other plants in the landscape plan.

Pruning time
The best time to carry out the main annual pruning of camellias is at the end of spring, when the new growth for the season is just beginning. This is because next season's flowerbuds form on this season's new growth. This new growth occurs in the spring and may vary according to the climate of your district. It is usually in the later part of the flowering season. Hence, if

40

Annual light pruning of a young bush, such as 'Daintiness' shown here, helps to maintain its shape.

A well-cared for mature camellia can make a spectacular specimen tree.

major pruning is delayed until after the new growth has appeared, most of next season's flowers will be lost. It is better to prune while the plant is still in flower, particularly when dealing with late-flowering varieties. Much of this pruning can be done when you are cutting flowers for the house or neighbours.

Young bushes

New camellias will not need pruning for the first two or three years, but it is better to start pruning well before the plant reaches the desired ultimate size. Light pruning each year will achieve a much better result than a particularly heavy pruning of a more mature, thick-branched and overcrowded camellia. Odd straggling branches can be pruned at any time, and it is wise to remove any branches too close to the ground because they rarely give perfect flowers, being prone to mud damage. Other branches can be shortened or removed to keep the desired shape and to open out the plant. As with roses, it is advisable to shorten the growth to a bud facing in a suitable direction for the next growth run.

When a camellia is grown in a restricted space, it is wise to allow one central trunk to develop and to remove or shorten any lateral branches which are thickening into heavy branches. This will give you an upright, lightly branched bush which is easy to deal with in the small garden.

Sanitary pruning

Sanitary pruning opens up and thins out a bush, allowing for light to penetrate the interior of the plant, which ensures good healthy growth. The flowers are produced mostly at the ends of branches and an unpruned camellia will have an overcrowded and heavily shaded interior where twigs and small branches will become unhealthy and eventually die. These overcrowded interiors are prone to infestation by scale and thrips. The Chinese say that a camellia should be pruned 'so that a bird may fly through it'.

With this lighter, more open growth pattern, spraying may be done more effectively, flowers will not damage so easily and the aesthetic result will be much more pleasing. This type of pruning may be done at any time of the year.

Pruning

pruning cut and bushy new growth appearing

When shortening a branch, ensure that you prune back to a leaf bud pointing in a desirable direction, just as you do with roses.

tip growth before pruning

Mature camellias

Any old cultivar which has outgrown its space in the garden and has an unsightly appearance can be pruned back drastically, preferably at spring's end. Even a mature *C. reticulata* may be reduced to bare main branches, resembling a hat-rack. I noted this happening in our local botanic gardens and thought how ugly the stumpy bare branches looked, but before many weeks had passed the whole bush was clothed with fresh new growth and after thinning in later years the bush was completely rejuvenated.

How to prune

Basic pruning

To open the plant for improved air circulation and light penetration, remove dead wood, all weak and spindly branches, and any branches crossing awkwardly over others. (See diagram above.) Individual pruners must work to achieve an aesthetically pleasing bush shape. It is a matter of personal preference.

A good manageable height for camellias in the home garden is somewhere between 1.5 and 2 m. At this size the tree can be cared for from ground level to the top and even the top-most blooms can be easily reached.

Finally, after pruning, to restore an even balance between the reduced foliage and the larger root system push a sharp spade into the ground around the plant to cut through some of the roots.

Leaf-bud thinning

During the growing season, and for some varieties again in the autumn, vegetative buds fatten and swell before exploding into leaf and developing into new branches. It is advisable to thin some of these by removing them with a gentle twist as soon as the growth pattern becomes apparent. This 'rubbing off' of the leaf buds is easy to do and will minimise later pruning. Remove only those that seem likely to grow awkwardly or to cause overcrowding.

Pruning hygiene

Sterilisation of secateurs with methylated spirits, particularly after pruning virus-infected camellias, is a recommended practice to prevent the spread of disease to healthy plants. Modern experts do not insist on the use of pruning paste or paint to seal pruning cuts, but individual gardeners often prefer to seal all cuts as an additional precaution against the invasion of diseases.

Chapter 7

PROPAGATION

'Lipstick'

A description of reproductive parts of the camellia flower will I hope aid the understanding of the techniques which follow. The camellia flower is capable of self-pollination because it is bisexual, i.e., the bloom contains both male and female reproductive organs (see diagram on page 45). The male organ is called the stamen. It consists of the anther, which holds grains of pollen, and a stalk known as a filament. The female organ is called the pistil. It consists of the stigma, which captures pollen, and the style and is connected to the ovary.

Pollination takes place when ripe pollen is transferred from an anther to a stigma. The stigma must be receptive, i.e., it secretes a sticky substance, and fertilisation occurs when the pollen grains, having reached the stigma, grow down the style to join with the ovule in the ovary. When the ovule has been fertilised, seeds develop in capsules or pods.

Saving seed

Collect seeds in the autumn when the seedpods have ripened and before they have split open and dispersed their contents around the base of the tree. New seedlings may take from two to eight years to produce their first flowers and it may be a further two or three years before the flowers become consistent in shape and form. These 'chance' seedlings will not produce plants identical to the parent plant, although they may be very similar. Only wild species, which are self-pollinated, will reproduce the same plant and flowers. Many seedlings will be uninteresting and only useful for understock, but you may be lucky and produce a distinctive new cultivar.

The anthers, stigma and style are clearly visible on *Camellia* 'Brilliant Butterfly'.

Seed sowing

Seed can be sown when fresh under the parent plant. Fairly extensive taproots will develop under the seedlings and these will need to be undercut in the open ground. Transplant the seedlings later. Seed which has been stored for later sowing will need to be soaked in warm water for 36 hours before sowing. Shaking the seed in a plastic bag with some Benlate powder will prevent it from rotting during the germination process.

Sphagnum moss can be used to germinate the seed initially. The developing taproot is then cut back and the seedling transplanted. In this way the formation of a strong, fibrous root system will be stimulated.

Darkness, warmth and moisture are the important requirements for the successful germination of seed.

Hybridising

Pollination by bees

Bees are the natural pollinators, carrying pollen from one plant to another on their legs as they search for honey. If pollination by bees has occurred and the seed is collected before it falls from the tree, the seed parent of the new plant will be known. The pollen parent or male donor, however, will be unknown and the only means of identifying and classifying the new plant as belonging to a particular species will be by making an educated guess guided by observation of leaf, flower and growth characteristics. Some excellent cultivars have come from these chance seedlings. One popular camellia of the present day, whose full parentage is not recorded, is 'Spring Festival', a *C. cuspidata* seedling.

Hand pollination

You control this means of pollination. It is an arranged marriage, with the pollinator deciding on the parents of the new camellia. Such controlled hybridising aims to change the growth habit, or improve the existing one, gain a new colour, create a new or improved flower form, gain perfume, and perhaps alter the size of the plant.

Hybridisers need to keep their specific goals in mind and choose the two parents carefully.

Hand pollination techniques

Dealing with the female or seed parent.
1. Seek out a bud on the female parent which is nearly ready to open and so has not yet been contaminated by any other pollen.
2. While the petals are developing colour and expanding, but before the stamens are exposed, carefully cut away the petals and also the anthers which produce the pollen.
3. Hold the bud firmly and with a razor blade or a sharp pair of scissors cut deeply enough around the entire flower, just above the green calyx, to remove the petals.
4. Remove all the stamens with tweezers or small scissors, being very careful not to damage the pistil and its stigma. This process is called 'emasculation' and it is necessary to allow the reproductive parts of the flower to be easily reached and also to prevent the possibility of self-pollination. (See diagram.)

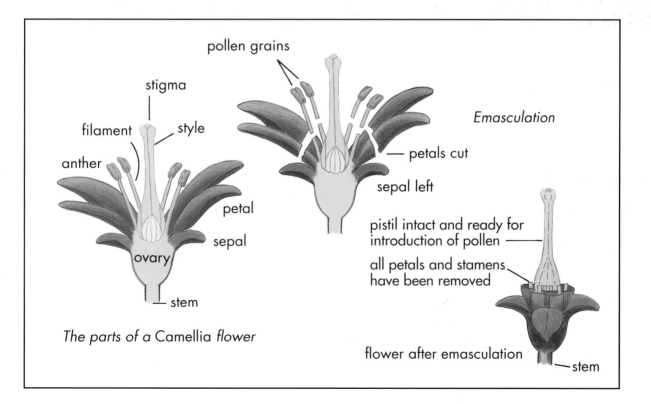

The parts of a Camellia *flower*

flower after emasculation

NB Identify the stigma carefully and be very sure to leave it intact. It is useful to practice dissection on a similar type of bloom to make sure that you can identify the reproductive organs correctly before you tackle the real thing.

5. Cover the prepared flower with a paper bag, to prevent chance pollination.

6. When the stigma becomes sticky, after three or four days (it will appear to have a globule of moisture on its tip), it will be receptive to pollen.

Preparation of the male or pollen parent.
As with the seed parent, just as the petals of the chosen male bloom are opening, protect the whole flower with a paper bag to prevent any pollen from an unknown source being deposited on it. Petals may be left on or removed as preferred. Only the stamens will be used.

The pollen, which is fluffy and bright yellow, should be collected and stored in the refrigerator in a small phial until you are ready to use it, because the stamens of the pollen-bearing male parent often reach maturity sooner than the stigma of the seed-bearing female parent.

The pollination routine

1. The ideal day temperature at which pollination should be carried out is at least 15°C. Transfer the pollen from one flower to the other with a small paintbrush, a fingertip or a matchstick. Remove the flowerhead from the male donor parent, and dust the pollen directly onto the sticky stigma of the female parent.

2. Alternatively, a few pollen-bearing stamens can be removed from the male donor, taken to the female parent and gently wiped onto the stigma.

3. Repeat the pollination process on at least two successive days, but remember to cover the flowerhead after each pollination.

4. Tie on a waterproof label, carefully noting the name of the seed parent and the pollen source. (The usual practice in describing the 'cross' is to write the name of the seed parent first, followed by the name of the pollen source. Add the date of the cross to the label and as much detail as you wish to help with any planned hybridisation programme. Record all the details a second time in a diary.)

5. Bag the pollinated flower immediately, to prevent any other pollen reaching the stigma.

Fertilisation will have been effected after about ten days, at which time it is possible to remove the bag.

After successful pollination, a seedpod will slowly develop and may be harvested the following autumn, before it fully splits open and ejects its seeds, i.e., before it is quite fully mature. Always label your seeds accurately. Sow them as described on page 44.

Camellias from cuttings

The simplest and most common method used to propagate camellias is by cuttings. Two types may be taken easily:

(a) Semi-mature wood The second and third months of summer, when the spring growth has ceased and the wood is suitably ripe and hard, are the best times of the year to take these cuttings.

(b) Mature wood Another good time to take cuttings is after the autumn growth has hardened in the second or third months of autumn.

Flowerbuds should be removed from any cuttings as soon as they are noticed.

C. japonica and C. sasanqua varieties strike well from cuttings, but those of C. reticulata do not and need to be grafted onto stronger rootstock. Many hybrids, including some crosses of C. reticulata, also strike readily from cuttings but, strangely, several C. japonica crosses can be difficult to strike and are delicate as young plants if they do

Reticulata hybrids, like 'Phyl Doak', do not strike well from cuttings and need to be grafted onto stronger rootstock.

take. One example is 'Ville de Nantes'.

Although early summer and autumn are the two optimum times for striking cuttings, they can be tried at any time of the year. The best time of the day to take cuttings is first thing in the morning and it is advisable to plant at once if possible. However, if necessary, the material can be kept for long periods in polythene bags made airtight by rubber bands and placed in the bottom of the refrigerator.

Dealing with semi-mature cuttings

1. Take cuttings with four or five nodes, i.e., 8–10 cm long, from upward-facing branches. Remove half of the top leaves of the cutting, to reduce loss of moisture by transpiration, and take off the lower leaves to provide a clear stem to poke into the potting mix.

2. Immerse the cutting completely in Captan or Benlate solution, and shake to remove surplus moisture.

3. Cut away a sliver of bark on one side of the stem and dip the wounded area in a rooting hormone powder or solution.

4. A recommended rooting mix is 75% pumice sand and 25% processed bark or peat, with no plant food added. (Use sterile material if possible, or sterilise the mix by pouring boiling water over it.) After filling a container with this mix, top off with coarse pumice sand.

5. With a dibbler, make vertical holes in the container of mix, and place a prepared cutting in each, being sure that the lower leaf left on the stem is above the surface of the mix. Lightly firm the mix round the cutting and water in gently.

6. To avoid moisture loss, and to maintain air humidity round the leaves, place the potted cutting inside a plastic bag or some other airtight container of a suitable size, or better still, put the container into a heated propagator. Keep the mix damp but not waterlogged while the cuttings are callousing and the roots are developing. Open the propagating case or plastic bag at least once a week to let air renew and freshen.

7. With bottom heat, roots may develop in six to eight weeks, but with no heat, rooting could take double or treble that time. When roots appear, allow more air in gradually by leaving the bag or propagator open for an hour or two extra each successive day, until at the end of a fortnight the plant is existing in a normal atmosphere. The container should be kept in good light, but not in full sun, at all times.

8. Use a very dilute liquid feed at this stage to encourage root extension.

Dealing with mature or hardwood cuttings

For cuttings taken in the autumn the procedure is as for semi-mature wood described above, but rooting time will be longer, and may be less successful.

Potting-on cuttings

When roots are well developed, pot-on cuttings into larger containers, using normal potting mix. Initially, after potting-on, increase the amount of shade a little to reduce transpiration loss while any damaged roots re-establish. Grow on your cuttings in semi-shade. After six months, examine the root system and if roots are visible all round, pot-on your developing cutting into the next size container, which should be at least 2 cm in diameter larger, and also deeper, than the existing one. At this stage,

Keep your camellia grafts in airtight conditions away from direct sunlight.

when potting-on, place a small stake alongside the stem to ensure upright growth.

Dealing with leaf-bud cuttings

This method should only be tried with bottom heat and a misting tent, otherwise the success rate is very low. However, leaf-bud cuttings can increase a variety quickly, provided really good propagating facilities are available.

1. In early summer, take a branchlet of semi-mature wood and cut it into sections, 2–3 cm long, each having a leaf-bud and a leaf. Cut the stem *above* the leaf *diagonally*, making the slope of the cut away from the bud eye, and then make a second *horizontal* cut about 1.5 cm *below* the bud eye.

2. 'Wound' each cutting by taking a thin sliver of bark wood off the side of the branchlet opposite the bud eye. Dip the wounded surface in your preferred rooting hormone solution or powder.

3. Fill small individual pots, or a tray, with simple potting mix, topped off with pumice sand. After making a vertical hole in the potting mix with a dibbler, place the leaf-bud cutting in this, being sure that the bud eye is just on or above the surface. Press the mix lightly around the cutting and water in.

4. After placing the container in a heated propagating box, wait for the elongation of the bud eye, which will indicate that a root system has been established. After six months, separate out and pot on each rudimentary plant.

Layering new plants

Layering in adjacent soil often occurs naturally, particularly with rhododendron and azalea bushes. It is the easiest, oldest and surest method of producing new plants identical to the parent plant.

Layering techniques for camellias

Although there are several variations on the

following method of layering, they all depend on the same basic principle of the partial arrest of the sap passing along the branch at a point where roots can be formed. An acute bend will normally cause sufficient interruption to the sap flow. Young wood, which can easily be brought down to ground level, is the best to use for layering, rather than older but larger branches. The soil into which the branch will be layered will need to be enriched with the addition of peat, compost, etc., and kept uniformally moist during the first summer after pegging down the branch.

When the soil has been prepared, the branch to be layered can be notched on its lower surface with a sloping cut half-way through, or it can be scraped so as to remove the bark on the underside. The prepared branch for layering is then brought down, buried shallowly in the rooting area and secured with a forked stick, wire pegs or a lump of rock. (Some gardeners maintain that the rock assists rooting.) Finally, the branch is tied as upright as possible to a cane or stick and left for two or three growing seasons.

After this time, a strong root system should have formed and the layer must be cut off from the original branch and left undisturbed for a while to get used to existing on its own root system. A few months later it can be lifted and shifted to a new position away from the parent plant, where it can grow to its full maturity. Some pinching and pruning may be necessary at this stage to form a well-shaped plant.

Air layering

Developed by the Chinese centuries ago, this method has a high success rate because the part to be rooted is still attached to the plant. The best time to try this method is in the first or second months of summer, when the bark is softer and can be easily removed. Choose a healthy camellia, selecting a branch with a good antler-like branchlet growth in its upper part, on the sheltered side of the bush if possible. The layer can be made 30 cm from the tip, but if a bigger plant is required, make the graft up to 60 cm from the tip.

1. With a sharp knife, girdle the stem with two parallel cuts around the branch, about 5 cm apart. Remove the ring of bark between them.

2. Next, scrape away the phloem and cambium layers which make up the green tissue between the bark and the wood, in order to retard healing. Dust the wound with a growth-promoting substance, then pack a handful of wet sphagnum moss in and around the cut surface and tie in place with string or waxed cotton thread, wound several times around the stem.

3. A square of black polythene (20 x 20 cm) can now be wrapped around the moss ball to prevent it drying out. This should be held in place temporarily with plastic twist ties, top and bottom, allowing for freer use of the operator's hands in the next stage of the process.

4. Tightly bind the top and bottom of the polythene to the branch with adhesive tape. It should be tight enough to prevent evaporation and the entry of rain. If the moss becomes waterlogged and thus oxygen-deficient, new roots cannot develop.

5. Moisten the root ball occasionally if it becomes too dry, as drying out of the moss will cause failure. From the cambial tissue exposed by the girdling, roots will develop and gradually extend through the moss ball.

6. After six months, begin to lightly squeeze the moss ball at intervals, repeating this until it feels firm, which will indicate that the layer is well rooted. It could take up to eight or nine months to reach this point. When the roots are evident, remove the ties and plastic covering, without breaking up the moss ball, and sever the branch below

the air layer, cutting off the stump as close to the roots as possible without damaging them. Reduce the foliage to keep it in proportion with the roots, then put the new plant in a container half filled with a standard potting mix. Fill the container and firm the mix. Place in a shady spot and keep moist.

7. Feed regularly with diluted liquid manure. Remove all flowerbuds during the first year of independance to encourage vegetative growth and the development of a strong, healthy plant.

Grafting

The process of encouraging material from one plant (the scion) to grow on the root system of another plant (the understock or rootstock) is called 'grafting'. This method is quicker than other techniques and useful for propagating *C. reticulata* varieties, which are difficult to grow from cuttings.

Terms used in grafting

Cambium layer This is the thin light-green-coloured layer of meristematic cells just beneath the bark.

Rootstock The stem and vigorous root system of a strongly growing plant from which the top growth has been removed. This understock will have the scion grafted onto it.

Scion The small shoot from the desirable variety, which will be united by grafting with the understock.

Cleft graft This is the most common form of grafting, where the top of the growing plant to become the rootstock is cut off 5–7 cm above the ground and the scion grafted onto the stump.

General grafting hints for camellias

The equipment necessary for grafting includes sharp secateurs, a very sharp knife, e.g., a pruning knife or a Stanley knife, and a small pruning saw, if the understock is thick enough to require it, also a small screwdriver. You will also need tying material, e.g., plastic string, raffia, rubber, adhesive tape or plaster, a covering for the completed graft, e.g., a plastic bag or a half-gallon glass jar with the bottom removed; a label to ensure the scion is correctly named; clean river sand; a hand lens; fungicide and rooting hormone; root stock in a container; a scion of a selected cultivar.

Have everything ready before you start and choose a sturdy shoot with a strong growth bud for the scion. If it cannot be grafted immediately, it can be sealed in an airtight, dampened plastic bag and kept cool in the refrigerator until required. The rootstock must be a vigorous, healthy plant with a strong root system which should be at least three years old. Don't use a plant recovering from the shock of recent transplanting or repotting. Ensure that both scion and rootstock are free of pests and diseases. If in doubt, spray with a fungicide or insecticide.

Suggestions for varieties which make good rootstocks: *C. sasanqua* 'Kanjiro' has good disease resistance and many hybridists have used it with worthwhile results. *C. reticulata* hybrids are also worth using as they are strong and fast growing. *C. japonica* hybrids are suitable as understock for other *C. japonica* varieties. However, it is inadvisable to graft a fast-growing *C. reticulata* hybrid onto a slow-growing *C. japonica* understock as this will result in uneven, ugly growth called 'bottle necking'.

The best time for grafting most varieties is in the last month of winter and the first month of spring, before new spring growth begins, although *C. reticulata* cultivars are best grafted in the last month of summer and the first month of autumn. However, it is possible to graft at any time you obtain a scion you want, provided that the plant is not in a period of active growth.

'Kanjiro' is a popular rootstock for grafting as it has good disease resistance.

Cleft grafting techniques

First prepare the rootstock, using a small saw or lopping shears by removing the top of the stock with an angled cut 5–10 cm above the ground. With a very sharp knife trim the cut cleanly to remove all ragged edges from the bark, then cut a cleft of the required length to take the scion in the centre of the stock. About 5 cm is a good average depth for this split. Hold the stump firmly with a pair of pliers during this operation, as there is always a danger that the knife could slip and injure an unprotected hand.

Next, shape the scion with the very sharp knife into the form of a wedge to fit into the cleft in the rootstock. The side of the wedge that will be on the inner side of the stock should be slightly thinner than the outer edge. Keep the cleft open by inserting the screwdriver and twisting it, and insert the scion into the cleft so that the cambium layers match evenly. The more contact there is between the cambium layers of scion and rootstock, the better the chance of a successful union.

Tie the graft firmly to keep it secure. Grafting paste spread over the joined area is a good insurance against disease entering but it is not essential. Cover the graft to exclude air, using a jar or a plastic bag. If using a glass jar, fill it with some sand around its base to ensure that it is airtight.

If the graft is in a position where it will be exposed to strong sunlight, it should be protected by a further covering to shield it from the sun. The graft should be clearly labelled with the name of the scion and any other information that will help to keep records and assist in the learning process. Inspect the graft weekly, returning it to the airtight conditions if all is well. Water sparingly only if the soil dries right out.

When you notice that the scion has begun to shoot and the union between the rootstock and the scion has taken, be very cautious about lifting the cover. This is a stage where it is all too easy to lose a graft through impatience. As growth increases, gradually allow the entry of more air, but if there are any signs of wilting make the cover airtight again until the graft recovers. (NB The new graft will collapse if the cover is removed before the graft union is fully developed.) If there is bleeding of the rootstock, watch for any fungal growth. Keep mopping up the moisture, and if there are signs of infections wipe the area with a fungicide solution. The tie can be removed when the graft has knitted or calloused over and the scion and the rootstock have clearly united.

Chapter 8

SEASONAL MAINTENANCE

'Tinkerbell' (left) and 'Gay Baby'

Spring

1. Wrench plants to be moved, and cut back the top growth when replanting to compensate for root loss.

2. The main pruning time is in the spring, after flowering. Large plants can be cut back if necessary to a stump a metre high, or even less. If using the hat-rack system for drastic pruning, do not do all your camellias in the same year as they will not flower for a year or two. (See Chapter 6 for general advice on pruning.)

3. After pruning, fertilise with your preferred mix.

4. Repot camellias in containers at this time. Root-prune and add slow-release fertiliser to the new mix. Root-prune plants in containers every few years. It is a good idea to prune two sides and repot in fresh mix. The alternate sides can be pruned the following year.

5. Spray for leaf roller.

Early summer

1. Try some aerial layers.

2. Watch for flying insects (grass-grub beetle) and other chewing pests. Spray if necessary or use other methods of prevention.

3. Mulch for the hotter months, using untreated sawdust, bark, compost, stones or live groundcovers. Remember to use lawn clippings only lightly.

4. Watering may be necessary at this time.

5. Constantly check grafts, cuttings, seeds and new plantings.

6. Remember to label everything accurately.

High summer

1. Camellias will need a lot of moisture at this time as new growth is hardening and buds are forming.

2. As well as soaking the soil, spray the foliage both on the surface and underside of the leaves.

3. If spraying for insects and not using an

oil spray, add a wetting agent. A foliar fertiliser may also be added to this.

4. Cultivate by lightly hoeing unmulched soil. A dust mulch will be created.

5. Stir the surface of an existing mulch to allow rain and water to penetrate.

6. Take cuttings of half-ripened wood of reticulata hybrids, non-reticulata hybrids and japonicas at this time for both propagation and grafting. Cuttings of some Kunming reticulata cultivars, such as 'Purple Gown', which are difficult to graft, are probably best taken at this time.

7. Cuttings may be planted in a light mix in containers in a heated bed or in a cutting bed in the garden.

Late summer

1. Be on the lookout for ripening seedpods and harvest them when cracks in the pod are noticed.

2. Do not allow seed to dry out but plant as soon as possible for the best results. (See Chapter 7.)

3. A little light pruning may be needed on some bushes. Remember to leave some branches with longer stems which will bear flowers for picking.

4. Disbudding may be done now, if preferred, especially where buds have formed in a cluster. A longer flowering season will be gained if some buds of all sizes are left. If some are left facing downwards they will better escape weather damage. The smaller-flowered species do not need to be disbudded as their charm lies in their floriferousness.

5. Plant out camellias layered in preceding years.

Autumn

1. This is a good time to plant out bushes that have been held over from last year or newly received from nursery orders. Make sure that the size of the planting hole is more than adequate for the size of the

A spring abundance of 'Donation'.

camellia and that some good soil mix is used around the roots.

2. Plant out, or pot on, any plants that you have propagated that are ready. Pot up aerial layers if they are showing healthy roots.

3. Pollinate sasanquas while they are in bloom. Have a definite objective in mind when undertaking hybridising projects.

Winter

1. Do some grafts if scions are available.

2. Continue to watch over earlier grafts and cuttings.

3. Watch for seed germination.

4. Late winter and early spring is a good time to start wrenching camellias you wish to move.

5. You can still plant camellias at the end of this season, and a useful tip is to place several thicknesses of newspaper in the base of the planting hole to retain moisture throughout the summer.

Chapter 9

CAMELLIAS IN LANDSCAPING

'Plantation Pink'

AS I indicated earlier, the introduction of new dwarf-growing species into breeding programmes has extended the range of camellia shapes and sizes available to the landscaper. The introduction of *C. saluenensis* into camellia-breeding programmes earlier this century provided much needed hardiness, a self-grooming habit, a longer flowering season and increased floriferousness. Now we have available camellias with a weeping habit, like 'Sweet Emily Kate', and plants with an upright, narrow, cyprus-like shape such as 'Spring Festival'. Dwarf camellias like 'Quintessence' and 'Baby Bear' make wonderful container plants. Others are good for hanging baskets, e.g., *C. transnokoensis,* and yet others, like 'Baby Bear', are excellent as bonsai subjects. Older varieties such as 'Barbara Clark' are ideal for providing a background to other shrubs, perennials and annuals in the open garden.

The sasanqua camellias begin blooming in late autumn and continue through the winter and early spring, so providing a long flowering season for the genus, and increasing its usefulness in the garden. Because they flower in winter and early spring, when many other plants are resting, these camellias have great value in the garden, and their average height of 2–3 m gives a substantial evergreen form to the back of the border. Wise choice of early- and late-flowering camellia varieties guarantees a long flowering season for six months of the year, and a careful selection of colours will ensure they harmonise with other plantings in the garden.

The natural habitat of the camellia is woodland, and this is the ideal site for

Deciduous trees are ideal companions for camellias, as their leaves give shelter from the hot summer sun and provide mulch in the autumn.

growing these plants where space is available. In a woodland garden they are protected from hot winds, strong summer sun and extreme winter cold by tall deciduous trees such as the deep-rooted oak, which makes an ideal shade tree and has leaves which break down into an acid mulch. A natural mulch is formed by the leaves of most shade trees and these conditions, while suiting camellias well, are also appreciated by rhododendrons and azaleas, which continue the pageant of colour later into the spring and early summer.

The choice of shade trees is wide, and some can provide a welcome colour contrast for the shiny deep green leaves of camellias. The *Prunus* species give colourful spring blossoms, as well as warm autumn tones in many cases, and *Magnolia* species and their hybrids, which come in evergreen and deciduous forms, are valuable smaller shade trees, especially with their often fragrant blossoms in shades of white, pink and purple. The *Cornus,* or dogwood, family from North America provides garden interest with their colourful, flower-like bracts in spring and their wealth of superb autumn colour. And don't look past the Japanese maples (*Acer palmatum*), which have been grown with camellias for centuries; with their finely textured leaves and wonderful diversity of colour, acers are perfect partners for the evergreen camellia.

Some larger colourful shade tress are *Robinia pseudoacacia* 'Frisia', the golden *Gleditsia,* the plum-coloured *Prunus* x *blireana* and *P. cerasifera,* and the silver and gold forms on *Acer negundo.* For silver tones try the weeping, silver-leaved pear, *Pyrus salicifolia* 'Pendula', or for a strong blue-grey element, the blue conifer *Picea pungens* 'Koster's Blue'. Experiment with form and colour to not only give shelter but to create exciting garden plantings.

Woodland bulbs like scilla, chionodoxa, grape hyacinth and fritillaries can create a carpet of seasonal colour and interest while being relatively easy to maintain beneath shrubs and large trees. Shade-loving, self-seeding annual and perennial groundcovers extend the display, as well as providing a useful mulch in the hotter months.

Camellias have been used as specimen trees for many centuries and many of those planted in gardens more than 100 years ago still thrive today. These earlier camellias were *C. japonica* hybrids, but in this century the tall and stately *C. reticulata* cultivars also make splendid specimen trees, and add the delights of their enormous brilliantly coloured and sometimes frilled blooms. The reticulatas will tolerate more sun than other varieties and so can be more suitable in an open position than the old japonicas. They are, however, less frost tolerant than *C.* x *williamsii* hybrids and are only suitable for more temperate areas or under glass in colder countries.

'Water Lily' as a standard.

Standards

To create a standard camellia the simplest method is to find a strong-growing plant with a straight stem and carefully prune away all growth from ground level to a suitable height, leaving a good balance of top growth in proportion to the thickness of the trunk. The top growth can now be pruned into a shape pleasing to your eye.

The alternative is to select a plant, once again with a strong, straight stem, but this time to graft a better variety on to it well above ground level. If weeping varieties are used for the scion, interesting and original effects can be achieved.

After-care, particularly in the first few years, is very important for a young standard camellia. Leaf buds will readily appear on the stock or trunk and must be promptly removed to prevent the formation of any branches below the head of foliage at the top. Extra care with regular pruning must be taken to preserve the attractive formal shape of the crown. Some lateral branches need to be pruned to a bud pointing in the desired direction, similar to the pruning of standard roses. Dominant and vigorous new growth must also be pinched back to ensure a regular outline.

Staking will help to preserve the straight upright line of the trunk. As the standard camellia grows taller, new growth on the leader will need to be tied to the stake at intervals until such time as it develops enough strength to stand unaided.

Varieties suitable for training as standards include 'Adorable', 'Anticipation', 'Ballet Queen', 'Contemplation', 'E. G. Waterhouse', 'Elegans Champagne', 'Flame' ('Moshio'), 'Hawaiian Bride', 'Jury's Yellow', 'Lady Loch', 'Rosiflora Cascade', 'San Dimas' and 'Spring Festival.'

Espaliers

Espaliers can be formed in various situations, but if a solid, sunny wall is chosen, a more heat-tolerant variety should be planted, although the effect of the heat will be reduced by fixing a close wooden trellis to the wall. Cultivars of *C. sasanqua* and *C. reticulata* are more heat-tolerant than *C. japonica* hybrids. 'Elegant Beauty' and 'Francie L' are both good subjects for growing on a warm wall.

It will be necessary to provide horizontal supports to which the branches can be tied, when growing camellias against any solid wall. These horizontal supports can be of wire, garden twine or slatted wood.

Various patterns of espalier are suited to camellias:

1. Branches can be trained horizontally from a main stem.
2. Branches may be trained to grow at an angle, also from one stem.
3. Two or more leaders may be trained from the main stem, with branches trained to go horizontally from these. (See diagram on page 58.)

When planting, take care to allow for good air circulation around and behind the branches. Tie the leader to the support as the plant grows and remove any superfluous growth with sharp, clean secateurs. Leaf buds and branches that grow downwards or towards the back should be rubbed off or cut off, as should any growth that interferes with the main pattern. Don't use wire for tying as it will cut into the growing branch, eventually ring-barking it and causing it to die. Use soft ties — old pantihose are ideal. Interlocking plastic mesh is good and can be bought by the metre, and rubber strips can also be used.

Main pruning should be done after flowering but just before the flush of spring growth, as is usual with all camellias. Time, patience and persistence are needed to develop a good espalier. Prepare the

A young sasanqua being started as an espalier.

ground well, as for hedge planting (see below), to ensure a lifetime of service from a strong espalier.

Healthy foliage and prolific flower production will be encouraged by regular feeding and watering. It is essential to ensure adequate moisture as the espalier is often at the foot of a wall which gets sparse rainwater. A good automatic system will take the worry out of watering for the busy gardener.

Varieties suitable for training as espaliers include 'Alpen Glo', 'Ariel's Song',

'Alpen Glow' trained along a fence.

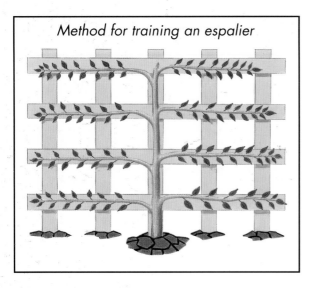

Method for training an espalier

An example of camellias used as a hedge.

'Bonanza', 'Cornish Snow', 'Coronation', 'Daintiness', 'Dream Girl', 'Dr Clifford Parks', 'Donation', 'Elegant Beauty', 'Flower Girl', 'Francie L','Jennifer Susan', 'Lucinda', 'Mine-no-Yuki', 'Nymph' and 'Plantation Pink'.

Hedges

Camellias can be clipped into trimmed hedges to form a solid evergreen barrier just as easily as any of the traditional hedging plants such as privet, holly or yew. The drawback with camellias is that the close clipping, to achieve a formal hedge shape, means the sacrifice of the flowerbuds, one of the chief glories of the garden. A flowering camellia hedge, which will grow closely upright for many years and need no clipping can be achieved, however, by planting carefully chosen varieties from a group of compact upright-growing cultivars, e.g., 'E. G. Waterhouse'. The minimal yearly trimming that is necessary should be carried out immediately after flowering. Generally plants should be planted no more than 80 cm apart, although a spacing of 1 m is preferable for the stronger-growing varieties, with much depending on position, soil conditions, and the exposure of the site to sun and wind.

Sasanqua cultivars make ideal hedging plants, with their bushy growth habit and their early flowering season. Because they start blooming in autumn and continue into early spring, a hedge of sasanqua camellias should be trimmed immediately after flowering and this routine should ensure that flowering is maintained the following year. The extra bushy growth of the sasanquas means that pruning need not be too heavy and so all flowers should not be lost.

The gardener can experiment with the composition of the colour scheme. An attractive one-colour hedge can be grown, but a tapestry of colours is possible when varieties that will flower at the same time are planted. Many *C. sasanqua* cultivars have the added bonus of perfume. 'Nicky Crisp' makes a most effective internal hedge about 1 m high. It has delectable pale pink flowers and its good bushy habit makes it neat and dense.

Combining in the one hedge camellias with similar growth habits and heights but with different flowering times can be an interesting experiment. Such a hedge will give you a succession of bloom for nearly six months of the year, if carefully planned.

A more relaxed-looking hedge can be composed of a mix of camellias with different growth habits. It will look more informal but still provide an effective barrier between one part of the garden and another.

Camellias suitable for formal or informal hedges: 'Anticipation', 'Barbara Clark', 'Berenice Boddy', 'Bettie Patricia', 'Dixie Knight', 'Donation', 'E. G. Waterhouse', 'Fircone', 'Firechief', 'Grand Slam', 'Lucinda', 'Nicky Crisp', 'Phyl Doak', 'Plantation Pink', 'Setsugekka', 'Shishi Gashira', 'Showa-no-Sakae' and 'Tootsie'.

Camellias in containers

Growing camellias in containers is an ancient practice. Records show that the Chinese have been growing these plants in pots for hundreds of years. With the increasing popularity of smaller town gardens, the potted camellia fills the need for an evergreen plant of beauty and substance, to give form to permanent plantings as well as to provide the added colour of its flowers in their season. Where space is at a premium, every plant must be carefully chosen so that it is useful or beautiful in more than one season. Camellias meet these criteria well, with their shiny leaves attractive throughout the year.

Camellias can be readily pruned to form standards, weeping standards, balls, columns, etc., which makes them useful both in the current revival of topiary in urban and courtyard gardens and also in larger gardens, on terraces and as potted and movable focal points. Mobility is their greatest advantage, and you can have one or another camellia in flower and taking centre stage in the garden for at least six months of the year. All this needs is care, when buying, to choose plants with different flowering times. In regions where climatic conditions are marginal for camellias, container-grown plants can be moved to

Well-grown container specimens.

the shelter of a greenhouse or conservatory to avoid severe winter cold, or into a shade-house to avoid blistering summer temperatures. The perfume of a scented variety in a container may be enjoyed to the full if it is moved to a doorway or near a much frequented pathway while it is in bloom.

By using containers, the camellia enthusiast is able to grow many more plants in a small garden by placing them on verandas and paths, or under carports. If the garden soil is an adverse one for camellias, a suitable one may be created in the containers, and their beauty enjoyed in this way.

Practical pointers for container selection

Nowadays there are a great many containers available: terracotta, metal, wood, plastic and concrete pots are all easy to come by and durable. Whatever type you choose, drainage holes are essential and must be of a reasonable size so that they will not become easily blocked. Broken terracotta pieces, stones or very coarse shingle placed over the drainage holes will help to prevent them from becoming blocked with hard-packed, impervious soil. Slightly raising containers off the ground is another means of ensuring free drainage if all the holes are at the bottom of the container. The sign of poor drainage is a yellowing of the camellia's leaves, which will eventually drop off. The condition must be corrected quickly by freeing the drainage holes or the plant will die. Waterlogging is more dangerous to camellias than dryness.

A container of suitable shape is most important. A pot which has a narrower neck than middle will be unsuitable, as it will be too difficult to remove the camellia for repotting. A container which is slightly tapered towards the bottom is the ideal shape; it is even preferable to a straight-sided one. It is better to start with a small container and pot-on the young camellia into a slightly larger one as it grows, until it has reached an appropriate size for its permanent container.

The soil mix used is very important. For the beginning gardener, a good commercial potting mix with a reputable brand name (there are several available from most garden centres) should be satisfactory. It should contain plenty of bark and organic material and also a slow-release fertiliser.

Planting

Before planting, soak your selected camellia overnight in a bucket of water. Thoroughly clean your container and ensure good drainage by placing broken crocks or stones over the drainage holes and covering them with gauze. Covering the bottom of the container with coarse, untreated pine bark to a depth of 2.5 cm will further aid good drainage.

Remove your new plant from its bag and inspect its roots, which should be light-coloured and vigorous looking. If they are dark and show any signs of decay, wash off all the old plant mixture with the hose and trim off any unsound-looking roots with

'Itty Bit' makes a charming container plant.

clean, sharp secateurs.

Place the plant in the container and carefully work the fresh potting mix round the roots, using your fingers and a blunt stick. The new potting mix should consolidate by about 5 cm and the container can now be topped up with a mulch of your choice. (Pine bark or pine needles are good.) Water in thoroughly.

Some reduction of the top growth by pruning will be necessary where there has been considerable reduction of the root system by trimming.

Aftercare

1. Take care to water containers regularly. A thorough soaking is more effective than frequent light sprinklings, and spraying the leaves in the cool of the evening will discourage insects, reduce temperatures and increase humidity.

2. Good regular feeding is essential, as frequent watering leaches nutrients from the soil. Slow-release fertilisers are beneficial, and regular foliar feeding with a weak solution of liquid fertiliser helps to ensure that pot-grown camellias are not underfed. Keep a careful watch to make sure that there is not a dangerous buildup of salts due to overfeeding with chemical fertilisers.

3. Container plants need regular pruning and disbudding, and repotting will be necessary as the plant outgrows its container and shows signs of stress through overcrowding of the roots.

4. Mulching will insulate the soil against extreme temperatures and may save the life of the plant. For year-round protection apply mulch in both spring and autumn, but be careful not to pile the mulch against the trunk of the camellia bush as the moisture and heat that are generated in the decomposition process could rot the bark, allowing disease organisms to enter, and possibly ring-barking the bush and causing its death. Mulch materials for containers are the same as those used for open-ground planting.

Camellia varieties suitable for containers: 'Adorable' (compact), 'Anticipation' (upright), 'Baby Bear' (compact), 'Baby Willow' (compact), 'Black Lace' (compact), 'Cinnamon Cindy' (upright), 'Demi-Tasse' (compact), 'Dresden China' (open or weeping), 'Elegans Champagne' (open or weeping), 'Elegans Supreme' (open or weeping), 'Garnet Gleam' (weeping or open), 'Glen Forty' (compact), 'Hawaii' (compact), 'Hawaiian Bride' (compact), 'Laurie Bray' (compact), 'Nicky Crisp' (compact), 'Night Rider' (compact), 'Prudence' (compact), 'Quintessence' (weeping or open), 'Rosiflora Cascade' (weeping or open), 'San Dimas' (compact), 'Scented Gem' (compact), 'Snippet' (compact), 'Snowdrop' (weeping or open), 'Spring Festival' (upright), 'Spring

'Sir Victor Davies' and *Corydalis* 'Pere David' in a container.

Mist' (weeping or open), 'Sugar Babe' (compact), 'Tammia' (compact), 'Tanya' (compact), 'Tiny Princess' (weeping or open), 'Tiny Star' (weeping or open), 'The Elf' (compact), 'Wirlinga Gem' (weeping or open).

Bonsai

The art of bonsai, which means 'plant in a tray', originated in China it is thought, about 500 BC. Later it spread to Japan, and the artistry of the bonsai growers in these two Asian countries ultimately inspired growers in the West. Ideally the bonsai camellia should be grown outdoors, provided it has shelter from the afternoon sun, which can cause root dehydration and foliage damage. Camellia seedlings from seed saved, or self-sown plants found under camellia trees, may be grown on until they have a strong leader with several branches suitable for planting in a bonsai pot and training. The tap root should be shortened to about half its length or less to encourage the growth of fibrous roots.

Another alternative is to choose a small but mature tree to train. Select a tree with a tapering trunk, making sure that the lower branches are strong and close to the base;

If you wish to try a bonsai camellia, 'Maroon and Gold' is an ideal variety.

the more branches the better so that there are plenty to be pruned away in order to train the tree into a pleasing shape. Match your container to the size and proportion of the tree and transfer the new bonsai into it, where it can be trained into shape for the next year. Refer to a detailed manual on bonsai if you intend to take up this fascinating hobby seriously.

Camellias suitable for bonsai: 'Alpen Glo', 'Baby Bear', 'Baby Willow', 'Bonsai Baby', 'Kuro Tsubaki', 'Mansize', 'Maroon and Gold', 'Mine-no-Yuki', 'Nicky Crisp', 'Snippet', 'Snowdrop', 'Spring Festival', 'Spring Mist', 'Tiny Star', 'Un-ryu' and 'Yuletide'.

Hanging baskets

Choosing the basket
The cultural methods for growing camellias in hanging baskets are the same as for other containers, but care must be taken to choose a big enough wire basket to allow space for growth of the camellia roots as the plant matures. A good starting size would be at least 30 cm wide at the top.

A light strong timber should be chosen for a wooden container 30 cm square at the top and approximately 15–17 cm deep. Whatever it is made of, the basket should not be too heavy to handle easily but should allow for adequate root expansion.

It is advisable to use some sort of liner to help retain moisture and to contain the potting mix.

Sphagnum moss is natural looking and adapts well to any shape. Soak it first in water containing liquid fertiliser, as it is easier to handle wet and will supply food as well. It needs to be packed in well to remain firm, and a chicken-wire lining will help to keep the moss in place and prevent birds from pulling strands out and leaving it looking raggy. There are other commercial liners available which are effective, some green and others hemp-coloured. Some are in round shapes to start with, and others need to be cut to fit your container. Experiment and choose the type that suits your basket shape and your personal preference.

Watering
The use of water-storage crystals (there are several good brands available) to help retain moisture in hanging baskets is recommended, as plentiful water and food make for a more effective display. If the crystals are soaked in water containing a liquid fertiliser, feeding and watering can be achieved in one easy operation. Liquid blood and bone, plus minerals and trace elements, is a good choice. Regular watering is crucial to the success of a hanging basket and an automatic watering system is ideal. There are various long-spouted watering devices on the market, and some gar-

Combine your camellias with something that will set off the blooms. Here *Omphalodes cappadocica* enhances the dusky blooms of 'Sir Victor Davis'.

deners resort to a system of pulleys which will lower the baskets easily for watering without too much strain on their muscles.

Pruning and training

More than usual care is needed in the pruning and training of camellias in hanging baskets. The main aim is to create a cascading effect by restricting growth upwards. Thus, when the basket is growing at above head height, the growth will be able to be fully appreciated from below. If a wire basket is used, it is easy to tie down the branches to the wires to encourage a cascading habit of growth. With a wooden container, it will be necessary to drive sta-

ples in low down on the basket and attach ties to these.

Copper wire, which is flexible, can be used to twist round a young branch and bend it into the desired position. It is wise to begin training the branches while they are young, when they will soon adapt to your requirements. The ties will need adjusting from time to time, and as the plant strengthens they can be dispensed with altogether. Unwanted growth should be removed regularly to achieve your desired shape.

Suitable varieties for hanging baskets:
C. lutchuensis, C. rosiflora, C. transnokoensis, C. tsai, 'Alpen Glo', 'Ariel's Song', 'Baby Willow', 'Cornish Snow', 'Dave's Weeper', 'Elegans', 'Elegans Champagne', 'Elegans Splendour', 'Elegans Supreme', 'Fragrant Pink', 'Fragrant Pink Improved', 'Quintessence', 'Rosiflora Cascade', 'Spring Mist', 'Tiny Princess', 'Wirlinga Belle' and 'Wirlinga Gem'.

'Rosiflora Cascade'

'Lemon Drop' (above) and 'Brushfield's Yellow' (opposite) are good creamy-yellow varieties to contrast with a strong foliage plant.

Window boxes

Many camellia varieties are suitable for window boxes. The fact that they are evergreen and have an early flowering time makes them appealing companion plants for spring bulbs, winter-flowering primulas, polyanthus and violas, as well as the flowering annuals and perennials of summer and autumn.

The possible combinations here are almost endless. Discerning gardeners learn to use ideas from many sources. Recently I helped to design and plant a garden for a wedding, using soft yellow, the colour of the bridesmaids' dresses, and blue, with plenty of white, lime-green and grey for softeners. On the morning of the wedding I delivered two hanging baskets that I had made, using the same colours, to hang under a pergola over a deck adjoining the house. I was fascinated to see the florists' arrangements — huge, rounded arrangements of 'Iceberg' roses with greenery. The genius touch, I thought, was the pale trails of gold and green ivy flowing and softening the edges. The whole effect was unstudied, pastoral and rather French-looking. Here was an idea for an arresting window box, using a white camellia and the green and gold ivy. The white could be a species, like *C. tsai* or *C. transokoensis,* or one of the new dwarf cultivars like 'Quintessence', 'Baby Willow' or 'Snowdrop'. The fluted white sasanqua 'Setsugekka', whose flowers do not mark easily in the rain, would be another possibility, as would the white miniature anemone form, 'Blondy'. From the white japonica cultivars the choice could include 'Polar Bear', 'Onetia Holland', 'Nuccio's Gem', and the beautiful New Zealand-bred 'Hawaiian Bride'.

One could reverse the colour combination, using a creamy-yellow camellia and a foliage plant like one of the *Parahebe* species, which has shiny, durable, fine green foliage and white flowers. (There are also blue- and lavender-flowered forms, which would be pretty in combination with soft pink camellias, as well as the creamy-yellow ones).

Some good cream and yellow varieties to use are 'Jury's Yellow', 'Brushfield's Yellow', 'Gwenneth Morey', 'Elegans Champagne' and 'Lemon Drop'.

Another variation of foliage in this yellow and white combination would be to use the lime-green *Helichrysum petiolare* falling gracefully over the edge of the window box. The addition of grey to the green, white and yellow scheme is attractive and softening, and the velvety, small, rounded leaves of the grey helichrysum species would be just as effective as a foil to the shiny green of the camellia leaves. Grey can also be very harmonious with pink-toned camellias, and a stronger, more masculine contrast could be achieved with bright red camellias and grey foliage.

The game of putting colours together gives endless pleasure and needs a book of its own. Study the camellia descriptions in the catalogue starting on page 68, and the illustrations, and make your own combinations for window boxes, remembering that it is the permanent, year-long component of the combination — the foliage — which is most important.

Care of camellias in window boxes

The same rules apply to planting, feeding and pruning camellias in window boxes, as do to those growing in containers. However, particular attention must be given to watering, as window boxes are not always in shade and so are subject to more evaporation loss than movable containers, which can be shifted into the shade in the hottest months of the year. It is a great advantage if some sort of automatic irrigation system can be arranged to ensure that watering is constant and reliable.

Camellias suitable for window boxes: *C. lutchuensis, C. rosiflora, C. transnokoensis, C. tsai,* 'Adorable', 'Alpen Glo', 'Annette Carol', 'Baby Bear', 'Baby Willow', 'Brushfield's Yellow', 'Cornish Snow', 'Dave's Weeper', 'Dresden China', 'Elegans', 'Elegans Champagne', 'Elegans Splendour', 'Elegans Supreme', 'Fragrant Pink Improved' and 'Gwenneth Morey'.

Camellias as specimen trees

Camellias make splendid specimen trees. There are varieties suitable for most garden design purposes. Those with flowers in the deeper colours stand up better to exposure to sun and frost than the paler pinks, whites and creams. *C. reticula* cultivars are tall and open growing, and stand more sun than japonicas, so are probably more suited to featuring as specimens in the landscape. Whatever the choice, do make sure that the colour of the selected camellia will harmonise with any background planting which may be seen within the scope of your eye. Other species blooming at the same time as the camellia must be considered carefully. Bear in mind, too, that staking is more important for specimen plants as they are usually exposed to more wind in a lawn or when planted alone and not supported or sheltered by associated shrubs and trees.

A single camellia planted at the back of a rockery composed of low-growing plants will give height to the garden picture. If there is space for a vista in the garden, a camellia makes a fine focal point to mark its end. Slender, columnar camellias such as 'Spring Festival' make handsome specimen trees. They can also be used effectively as accent plants on each side of a path or doorway. Similarly, a weeping standard camellia can make a beautiful specimen subject in a lawn, at the end of a vista or to finish off a path.

Suitable varieties for specimen trees: 'Anticipation', 'Aztec', 'Berenice Boddy', 'Bob Hope', 'Buddha', 'Captain Rawes', 'Crimson Robe', 'Donation', 'Dr Clifford Parks', 'El Dorado', 'Elsie Jury', 'Francie L', 'Guilio Nuccio', 'Harold L Paige', 'Hulyn Smith', 'Jubilation', 'Lasca Beauty', 'Lisa Gael' (narrow, upright), 'Mary Phoebe Taylor', 'Midnight', 'Moshio' ('Flame'), 'San Dimas', 'Spring Festival' (narrow, upright),

'Terrell Weaver', 'Valentine Day', 'Wildfire', 'William Hertrich' and 'Woodford Harrison'.

Camellias as groundcover

With the introduction into modern breeding programmes of the new dwarf-growing species, those with spreading habit, and the weeping varieties, entirely new possibilities emerged for the landscape use of camellias. One important alternative role for the camellia was as a groundcover plant. New hybrids such as 'Baby Bear', 'Itty Bit', 'Quintessence' and 'Nicky Crisp' have different attributes altogether from the old japonicas, which used to be thought of as the typical camellia. Some new hybrids have a sprawling growth habit, others are particularly slow growing, with some never reaching higher than 1 m. Others are particularly bushy. Most will train and prune readily into a desired shape. These qualities make them eminently suitable for planting on banks, at the edge of raised beds, on slopes, and to clothe the ground under deciduous shade trees in a woodland.

They can be used, too, in a purely evergreen landscape composition, under either standard camellias or the taller, more tree-like *C. reticulata* hybrids, which are more sun-tolerant than some of the old japonica crosses. One could use a sasanqua standard, which would start blooming in the autumn, underplanted with a spring-flowering, low-growing groundcover such as 'Nicky Crisp' or 'Quintessence'. This would give flower colour almost continuously from autumn to late spring, and the possibilities of summer colour combinations among the evergreen background of the summer-dormant camellias are endless. Try fuchsias, standard or bush, or both. Plant a tapestry of varied hostas whose lush leaves appear in mid spring and delight us till the frosts. Plant parahebes with blue, white or soft mauve blooms which have smaller, more finely textured evergreen leaves and enjoy some shade while they flower continuously during the summer. The colour possibilities are exciting. Imagine a sparkling bank of the ruby-coloured sasanqua 'Yuletide' warming the winter scene, or a delicate apple-blossom-coloured bank of the sprawling 'Rosiflora Cascade', with its masses of rosebud-shaped small flowers in the spring.

Camellias suitable for bank planting (these varieties hang their heads so look good when seen from below): 'Elegans Champagne', 'Elegans Splendour', 'Elegans Supreme', 'Tiffany'.

More open-growing groundcover varieties: 'Alpen Glo', 'Ariel's Song', 'Cornish Snow', 'Dave's Weeper', 'Elegant Beauty', 'Fairy Wand', 'Jennifer Susan', 'Lucinda', 'Nymph', 'Plantation Pink', 'Setsugekka' and 'Sparkling Burgundy'.

Some newer hybrids, such as 'Itty Bit' shown here, can be used as a groundcover.

'Snippet' is a slower growing groundcover variety.

Slow-growing groundcovers: 'Baby Bear', 'Baby Willow', 'Bonsai Baby', 'Itty Bit', 'Nicky Crisp', 'Snippet' and 'The Elf'.

Camellias with foliage interest

One of the most appealing attributes of the camellia is its shining evergreen foliage. It pays to remember and utilise some of the interesting variations in camellia leaf shape, texture, size and colour when composing garden pictures.

The sasanqua camellia 'Yuletide' is very compact and eminently suitable for shaping as a topiary subject. 'Lois Shinault' has bronze new growth in the spring, and the similarly coloured new growth of 'Wirlinga Princess' has a pinkish cast to it. 'Fragrant Pink Improved' also has gleaming bronze new growth. Watch for more varieties with distinctive spring colour at nurseries and on garden visits at this time of year.

There are several camellias which have variegated leaves. They can brighten up a gloomy corner and are well worth watching for. 'Golden Spangles' has a yellow centre with green edges whereas 'Reigyoku' has glossy green foliage with a blotch of pink in the centre when young, which changes to yellow on maturity. Camellias with the name 'Benten', e.g., 'White Doves Benten', are noted for their brilliantly coloured green leaves, evenly edged with creamy-yellow, white or pale green. They are worth asking for.

There are several interesting variations of leaf shape to bring interest to the camellia plantings in your garden. 'Hakuhan-kujaku' ('White Spotted Peacock') has distinctively curled leaves and flowers at the stem tips. 'Holly Bright', as its name implies, has unusual, crinkled, holly-like foliage, as has 'Lady Vansittart'. 'Kingyo-tsubaki' ('Mermaid' or 'Quercifolia') has very different 'fishtail' leaves. 'Sakura-ba-tsubaki' has light-green upward-curling leaves, with very deep serrations, almost a fringed appearance, while 'Un-ryu' has an unusual zigzag growth pattern on each leaf. Add to these leaf differences the long, narrow, pointed leaves of *C. tsai*, the small pointed leaves of *C. lutchuensis*, the often variegated leaves of *C. rosiflora*, the leaves of *C. salicifolia*, which open pink and change through bronze to green, and you will realise that the species camellias also have a wealth of interest in their diverse foliage. Do, then, consider foliage very carefully when making camellia purchases, and have a colourful garden year round.

Chapter 10
SELECTION OF
POPULAR VARIETIES

'Alpen Glow'

In compiling this list I have used information from camellia nursery catalogues, *Camellia Nomenclature* (the official nomenclature book of the American Camellia Society) and reference books mentioned in the bibliography. The origins of the camellia's breeding, when known, is indicated by the following initials:

J   *C. japonica* or cultivar
R   *C. reticulata* or cultivar
S   *C. sasanqua* or cultivar
W   *C.* x *williamsii* hybrid

Flowering times are indicated by:

VE   very early (late winter, early spring)
E   early spring
M   mid spring
L   late spring
VL   very late (into early summer)

Each description mentions the country of origin when known and the year of registration, or the year the cultivar became available commercially or was introduced to the West from Asia. Flower colour, size and form are included, and, where relevant, the growth habit is also described.

'Adolphe Audusson' Dark red. Large semi-double. Average compact growth. France, 1877. J, M
'Adorable' Bright pink. Medium formal double. Compact upright growth. Australia, 1979. *C. pitardii* seedling, M/L
'Akashi-gata' ('Lady Clare') Deep pink. Large, semi-double. Vigorous bushy growth. Japan, 1887. J, E/M
'Alba Plena' White. Medium formal double. Slow bushy growth. China, 1792. J, E
'Alpen Glo' Two shades of pink. Miniature

'Amazing Graces'

'Anticipation'

single to semi-double. Open upright growth. Australia, 1985. 'Snowdrop' seedling, M

'Amazing Graces' Blush-pink shading to deeper pink at edge. Small formal double with swirled inner petals. Average open upright growth. USA, 1979. J, M

'Annette Caroll' Pale pink. Small informal double to peony form. Tall open growth. Australia, 1981. *C. pitardii* seedling, M/L

'Anticipation' Deep rose. Large peony form. Upright growth. New Zealand, 1962. W, M

'Ariel's Song' Miniature single. Average open upright growth. New Zealand, 1990. *C. fraterna* hybrid, M/L

'Aspasia Macarthur' White to cream-white with a few rose-red lines and dashes. Medium full peony form. Slow, upright growth. Australia, 1850. J, E/M

'Aztec' Deep rose red. Very large, semi-double with irregular petals to loose peony form. Average open upright growth. USA, 1971. R, E/L

'Baby Bear' Light pink to white. Miniature single. Dwarf, compact growth. New Zealand, 1976. *C. rosiflora* hybrid, M

'Baby Willow' White miniature single. Average, dwarf, weeping growth. New Zealand, 1983. *C. rosiflora* seedling, M

'Ballet Dancer' Cream, shading to coral pink at edge. Medium full peony form

with mixed petals and petaloids. Average compact upright growth. USA, 1960. J, E/L

'Ballet Queen' Salmon pink. Large peony form. Average growth. New Zealand, 1975. W, M/L

'Barbara Clark' Rose pink. Medium semi-double. Vigorous compact upright growth. New Zealand, 1958. R, E/L

'Bellbird' Rose pink. Small single of bell shape. Vigorous spreading growth. Australia, 1970. 'Cornish Snow' seedling, M

'Berenice Boddy' Light pink with deep pink under petals. Medium semi-double. Vigorous upright growth. USA, 1946. M

'Bettie Patricia' Persian rose. Medium rose form double. Vigorous upright growth. S, E

'Barbara Clark'

'Bob Hope'

'Betty Sheffield Supreme' (sport of 'Betty Sheffield') White with deep-pink to red border on each petal. USA, 1960. J, M

'Black Lace' Dark velvet-red. Medium rose-form to formal double with incurved petals. Compact upright growth. USA, 1968. R, M/L

'Blondy' White. Miniature anemone form. Open upright growth. Australia, 1986. J, E/M

'Blue Danube' Rose-lavender. Medium peony form. Vigorous upright growth. USA, 1960. J, M

'Bob Hope' Black-red. Large semi-double with irregular petals. Slow compact growth. USA, 1972. M

'Bob's Tinsie' Brilliant red. Miniature to small anemone form. Average compact upright growth. USA, 1962. J, M

'Bokuhan' ('Tinsie') Red outer-guard petals and white peony centre. Miniature anemone form. Average compact upright growth. Japan to USA. J, E/M

'Bonanza' Deep red. Medium peony form. Can be trained as a standard on a single stem without grafting. S, E

'Bonsai Baby' Deep red. Small formal to rose-form double. Low spreading growth. *C. hiemalis* hybrid, E

'Bow Bells' Pink single with bell-shaped flowers. Compact bushy growth suitable for a shady wall. Can bloom from mid winter to late spring. England, 1925. W, VE-L

'Brushfield's Yellow' Antique white guard petals surround double centre of lightly ruffled pale primrose-yellow petaloids. Medium anemone form. Vigorous compact columnar growth. *See also* 'Gwenneth Morey', which is similar but not identical. Australia, 1968. J, M/L

'Buddha' Rose pink. Very large semi-double with irregular upright wavy petals. Vigorous upright growth. China to USA, 1950. R, M

'Buttons and Bows' Light pink shading deeper at edge. Small formal double. Average compact growth. USA, 1985. *Saluenensis* hybrid, E/M

'C. M. Hovey' Dark red. Medium formal double. Medium slender upright growth. USA, 1853. J, L

'Can Can'

Above: 'Cinnamon Cindy'

Below: 'Christmas Daffodil'

'C. M. Wilson' (A sport of 'Elegans') Light pink. Large to very large anemone form. Slow spreading growth. USA, 1949. E/M.

'Can Can' (Sport of 'Lady Loch') Pale pink with darker veining and petal edges. Medium growth. Australia, 1961. J, E/M

'Captain Rawes' Carmine rose pink. Very large semi-double with irregular petals. Average open growth. China to England, 1820. R, L

'Carter's Sunburst' Pale pink striped or marked deeper pink. Large to very large semi-double to peony form to formal double. Average spreading growth. USA, 1959. J, E/L

'China Clay' White. Medium semi-double. Open growth. England, 1972. W, M/L

'Cho Cho San' Light pink. Medium, semi-double to anemone form. Average compact growth. Japan to USA, 1936. J, M

'Christmas Daffodil' White tinged blush-pink at petal tips. Small anemone form. Vigorous compact growth. USA, 1971. J, E/M

'Cinnamon Cindy' Rose pink with white centre petaloids. Miniature peony form. USA, 1973. *C. lutchuensis* hybrid, E/M

'Confucious' Orchid pink. Large semi-double, with high centre and intermingled petaloids and stamens in centre. Average compact upright growth. China to USA, 1950. R, M

'Contemplation' Lavender pink. Medium semi-double with occasional petaloids. Slow compact growth. New Zealand, 1975. J, M/L

'Cornish Snow' White with occasional pink blush. Small single. Open upright growth. England, 1950. *C. saluensis* hybrid, M.

'Cornish Spring' Rich pink. Single. Vigorous upright bushy growth. England, 1950. *C. cuspidata* hybrid, M/L

Above: 'Cotton Candy'

Below: 'Confucious'

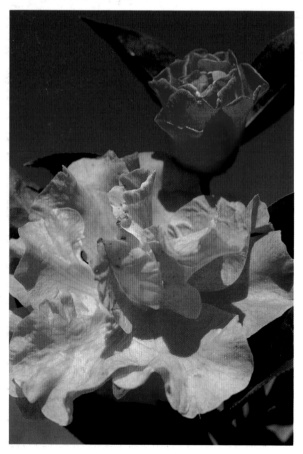

'Coronation' White. Very large semi-double. Vigorous open spreading growth. USA, 1954. J, M

'Cotton Candy' Clear pink. Medium semi-double with ruffled petals. S, E

'Crimson Robe' Carmine red. Very large, semi-double with wavy, crinkled, crepe-textured petals. Vigorous spreading growth. China to USA, 1948. R, M

'Dahlohnega' Canary yellow. Small to medium formal double. Slow open upright growth. USA, 1986. J, M

'Daintiness' Salmon pink. Large semi-double. Average open growth. New Zealand, 1965. W, M

'Daitarin' Light rose pink. Large single with mass of petaloids in centre. Vigorous upright growth. Japan, 1941. J, E

Above: 'Daintiness'

Above: 'Daitarin'

'Debbie' Clear rose to deep cyclamen pink. Medium to large peony form. New Zealand, 1965. W, M

Below: 'Debbie'

'Debutante' Light pink. Medium full peony form. Vigorous upright growth. USA, early 1900s. J, E/M

'Demi-Tasse' Peach-blossom pink. Small to medium semi-double hose-in-hose form with row of petaloids between petals. Vigorous upright growth. USA, 1962. J, M

'Desire' Pale pink edged deep pink. Medium formal double. Vigorous, compact upright growth. USA, 1977. J, M

Below: 'Desire'

'Dixie Knight'

'Dr Clifford Parks'

'Dixie Knight' Deep red. Medium loose peony form with irregular petals. Vigorous upright growth. USA, 1955. M/L

'Donation' Orchid pink. Large semi-double. Compact upright growth. United Kingdom, 1941. W, M

'Dr Clifford Parks' Red with orange cast. Very large semi-double loose peony form, full peony form or anemone form. Vigorous growth. USA, 1971. R, M

'Dr Tinsley' Very pale pink at base, shading to deep pink at edge with reverse side flesh-pink. Medium semi-double. Compact upright growth. USA, 1949. J, M

'Drama Girl' Deep salmon to rose-pink. Very large semi-double. Vigorous open pendulous growth. USA, 1950. J, M

'Dream Boat' Bright pink with lavender cast. Large formal double with incurved petals. Average open upright growth. New Zealand, 1976. W, M

'Dream Girl' Salmon pink. Large to very large semi-double with fluted petals. Vigorous upright growth. USA, 1965. R, E

'Dresden China' Pale pink. Large peony form. Slow spreading upright growth. New Zealand, 1980. W, M/L

'E. G. Waterhouse' Light pink. Medium formal double. Vigorous upright growth. Australia, 1954. W, M/L

'El Dorado' Light pink. Large full peony form. Average spreading growth. USA, 1967. J, M

'Elegans' Rose pink with centre petaloids often spotted white. Large to very large anemone form. Slow spreading growth. England, 1831. J, E/M

'Elegans Champagne' (sport of 'Elegans Splendour') White with cream centre petaloids, with pink occasionally at base of petals. Large to very large. USA, 1975. J, E/M

'Elegans Splendour' (sport of 'C. M. Wilson') Light pink edged white with deep petal serrations. Foliage and type of flowers similar to 'Elegans Supreme'. Large to very large. USA, 1969. J, E/M

'Elegans Supreme' (sport of Chandler's 'Elegans') Rose pink with very deep petal serrations. Large to very large. USA, 1960. J, M

'Elegant Beauty' Deep rose. Large anemone form. Open upright growth, new growth is reddish-coloured. New Zealand, 1962. W, M

'Elfin Rose' Rose pink. Azalea-form double. Tall and slender, ideal to grow where space is limited. *C. hiemalis* cultivar, E

'Elsie Jury' Clear pink with shaded orchid-pink undertones. Large full peony form. Average open growth. New Zealand, 1964. W, M/L

Above: 'Dream Girl'

Below: 'E.G. Waterhouse'

'E. T. R. Carlyon' White. Medium semi-double to rose-form double. Vigorous upright growth. England, 1972. W, M/L

'Fairy Wand' Bright rose-red. Miniature semi-double. Average upright growth. New Zealand, 1982. W, M

'Flame' ('Moshio') Deep red. Medium semi-double. Upright compact growth. Japan to Australia. J, M

'Flower Girl' Pink. Large to very large semi-double to peony form. Vigorous upright growth. USA, 1965. R, M/L

'Fircone' Blood-red. Miniature, semi-double, similar to a fircone. Vigorous bushy growth. USA, 1950. J, M

'Fire Chief' Deep red. Large semi-double to peony form. Average spreading upright growth. USA, 1963. R, L

'Fragrant Pink' Deep pink. Miniature peony form. Fragrant. Average spreading growth. USA, 1968. *C. lutchuensis* hybrid, E/L

'Fragrant Pink Improved' Fragrant-flow-ered form of 'Fragrant Pink'. Has good bronze spring foliage. USA, 1975. E/L

'Francie L' Rose pink. Very large semi-double with irregular wavy petals. USA, 1964. R, M

'Garnet Gleam' Vivid garnet-red. Small to medium, trumpet-shaped single with pink filaments and gold stamens. Slow spreading growth. New Zealand, 1980. *C. pitardii* hybrid, E/L

'Gay Baby' Deep orchid-pink. Miniature semi-double. Open upright growth. New Zealand, 1978. M

'Gay Border' White with broad pink border. S, E

'Gay Pixie' Light orchid-pink with deeper pink stripes. Large peony form. Open upright growth. Australia, 1979. *C. pitardii* seedling, M/L

'Gay Sue' White with cream anthers. Medium semi-double with frilled petals. S, E

'Ginger' Ivory-white. Miniature full peony form. Average upright growth. USA, 1958. J, M/L

'Glen Forty' Deep red. Medium to large formal to rose-form double. Slow compact upright growth. USA, 1942. J, E

'Glowing Embers' Red. Very large semi-double to loose peony form. Average open upright growth. New Zealand, 1976. R, E

'Golden Spangles' Variegated leaf form of 'Mary Christian'. Has leaves that are yellow in the centre with green edges. Small single phlox-pink flowers. Open upright growth. England, 1957. W, E/M

'Grand Prix' Brilliant red. Very large semi-double with irregular petals. Vigorous upright growth. USA, 1968. J, M

'Grand Slam' Brilliant dark red. Large to very large semi-double to anemone form. Vigorous open upright growth. USA, 1962. J, M

'Fragrant Pink Improved'

Above: 'Grand Slam'

Above: 'Guillio Nuccio' Variegated

Below: 'Gwenneth Morey'

'Grand Sultan' Dark red. Large semi-double to formal double. Slow open growth. Italy to Belgium, 1849. J, M/L

'Guest of Honor' Salmon pink. Large to very large semi-double to loose peony form. Vigorous compact upright growth. USA, 1955. J, M

'Guillio Nuccio' Coral rose pink. Large to very large semi-double with irregular petals. Vigorous upright growth. USA, 1956. J, M

'Gwenneth Morey' White outer-guard petals and deep cream to primrose-yellow petaloids. Medium anemone form. Average upright growth. Australia, 1965. J, M/L

'Hakuhan-kujaku' ('White Spotted Peacock') The peacock camellia. Red mottled white. Small single with slender tubular petals, has distinctively curled leaves and flowers at the stem tips. Medium semi-cascading growth. Japan, 1956. J, M/L

'Harold L. Paige' Bright red. Very large rose form double to peony form. Vigorous spreading growth. USA, 1972. R, E/M

'Harry Cave' Deep scarlet red. Medium semi-double. Slow compact growth. New Zealand, 1991. J, E/M

Above: 'Hawaiian Bride'

Below: 'High Fragrance'

'Hawaii' Pale pink. Medium to large peony form with fimbriated petals. Slow spreading growth. USA, 1961. J, E/M

'Hawaiian Bride' (sport of 'Hawaii') Pure white. Medium fully peony form with fimbriated petals and firm texture. Vigorous upright growth. New Zealand, 1992. J, M/L

'High Fragrance' Pale ivory-pink with deeper pink shading at edge. Medium peony form. Vigorous open growth. New Zealand, 1986. J, M

'Hiraethlyn' Pink shading to darker pink. Medium single of funnel form. Vigorous compact upright growth. England, 1950. W, M/L

'Holly Bright' Glossy salmon-red. Large semi-double with crepe petals. Has unusual crinkled holly-like foliage. Average compact upright growth. USA, 1985. J, M

'Howard Asper' Salmon-pink. Very large peony form with loose upright petals. Very large heavy foliage. Vigorous spreading upright growth. USA, 1963. R, M/L

'Hulyn Smith' Soft pink. Large semi-double. Average upright growth. USA, 1979. R, M/L

Below: 'Hulyn Smith'

'Itty Bit'

'Inspiration' Phlox pink, brighter than 'Donation'. Medium semi-double. Very floriferous. England, 1954. R, E/L

'Isaribi' Rose pink. Miniature semi-double. Vigorous compact upright growth. USA, 1981. J, M/L

'Itty Bit' Soft pink. Miniature anemone form. Slow spreading growth. New Zealand, 1984. *C. saluenensis* hybrid, M

'Jamie' Vivid red. Medium semi-double of

hose-in-hose form. Australia, 1968. W, M.
'J. C. Williams' Phlox-pink. Medium single. Vigorous upright growth. England, 1940. W, E/L

'Jean Clere' (sport of 'Aspasia Macarthur') Red with narrow band of white around the edge. Medium full peony form. Medium growth. New Zealand, 1969. J, M

'Jennifer Susan' Pale pink. Medium rose-form double with curled petals. S, E

'Joan Trehane' Rose pink. Medium rose-form to formal double. Spreading growth. England, 1980. W, L

'Jubilation' Pink with occasional deeper pink fleck. Large to very large rose-form double. Upright growth. New Zealand, 1978. M/L

'Julia Hamiter' Delicate blush-pink to white. Medium semi-double to rose-form double. Average compact growth. USA, 1964. Seedling of 'Donation', M

'Jamie'

'Jury's Yellow' White with creamy-yellow petaloids. Medium anemone form. Average compact upright growth. New Zealand, 1983. W, E/L

'Kanjiro' ('Hiryu') Rose-pink shading to rose-red at the edges of petals. Large semi-double. Tall and bushy. A strong grower and an excellent performer. A good red-flowered hedger. *C. vernalis,* E

'Kathryn Funari' Deep veined pink. Large formal double. Average growth. USA, 1975. J, E

'Kewpie Doll' Chalky light pink. Miniature anemone form with high petaloid centre. Vigorous bushy upright growth. USA, 1971. J, M

'Kick Off' Pale pink marked deep pink. Large to very large loose peony form. Vigorous upright growth. USA, 1962. J, M

'Kingyo-tsubaki' ('Mermaid' or 'Quercifolia') The goldfish camellia. Strawberry-icecream to scarlet coloured. Large single. Distinctive for its fishtail leaves. E/L

'Kramer's Supreme' Turkey red. Large to very large full peony form. Fragrant. Vigorous compact upright growth. USA, 1957. J, M

'Kramer's Supreme'

'Lady Loch'

'K. Sawada' White. Large formal to rose-form double. Vigorous semi-upright growth. M

'Kuro Tsubaki' Black-red with red stamens. Small semi-double. Average compact growth. Japan, 1896. J, L

'Lady Clare' Deep pink. Large semi-double. Vigorous bushy growth. Japan to Europe, 1887. J, E/M.

'Lady Loch' (sport of 'Aspasia Macarthur') Light pink, sometimes veined deeper pink and edged white. Australia, 1898. J, M

'Lady Vansittart' White striped rose-pink. Medium semi-double with broad, wavy-edged petals. Slow, bushy growth. Japan to England, 1887. J, M/L

'Lasca Beauty' Soft pink. Very large semi-double with heavy textured thick petals. Vigorous open upright growth. USA, 1973. R, M

'Laura Boscawen' Deep rose-pink. Anemone form. Very floriferous. Bushy growth. England, 1985. W, M

'Laurie Bray' Soft pink. Medium to large semi-double with spaced and ruffled petals. Upright growth. Australia, 1955. J, M

'Lemon Drop' White with lemon centre. Miniature rose-form double to anemone form. Average dense upright growth. USA, 1981. J, M

'Leonard Messel' Rose. Large semi-double. England, 1958. R, E/L (long flowering season)

'Leonora Novick' White. Large to very large loose peony form. Average upright growth. USA, 1968. J, E/M

'Lila Naff' Silvery pink. Large semi-double with wide petals. Vigorous compact upright growth. USA, 1967. R, M.

'Lipstick' Waxy red outer petals and white petaloids with a touch of red on their edges and base. Miniature anemone form. Small foliage and slow growth. A smaller grower than 'Tinsie', useful for the town garden. USA, 1981. J, M

Below: 'Laurie Bray'   Above: 'Lasca Beauty'

'Man Size'

'Lisa Gael' Rose pink. Large rose-form double. Compact upright growth. New Zealand, 1967. R, M

'Little Lavender' Lavender pink. Miniature anemone form. Vigorous compact upright growth. USA, 1965. J, M

'Little Slam' Rich red. Miniature full peony form. Average compact upright growth. USA, 1969. J, E/M

'Mark Alan'

'Lois Shinault' Orchid pink, lighter pink in centre. Very large semi-double with irregular petals ruffled on edges and upright central petals. Bronze new foliage. Average spreading growth. USA, 1973. R, E/M

'Lovelight' White. Large semi-double with heavy petals. Vigorous upright growth. USA, 1960. J, M

'Lucinda' Pink. Medium peony form. S, E

'Madame Picouline' ('Akaroa Rouge') Bright cherry red. Medium full peony form. New Zealand, 1970. J, M

'Man Size' White. Miniature anemone form. Average upright growth. USA, 1971. J, M/L

'Margaret Davis' (sport of 'Aspasia Macarthur') White to creamy-white with few rose-red lines and dashed and edged bright vermillion. Medium informal double. Strong upright growth. Australia, 1961. J, M/L

'Margaret Hilford' Deep red. Very large semi-double. Vigorous open upright growth. New Zealand, 1980. R, E

'Mark Alan' Wine red. Large semi-double to loose peony form with smallish petals. Average compact upright growth. USA, 1958. J, E/M

'Maroon and Gold' Maroon. Small to medium loose peony form with golden stamens. Vigorous upright growth. USA, 1961. J, M/L

'Mary Phoebe Taylor' Light rose pink. Very large peony form. Average open upright growth. New Zealand, 1975. *C. saluenensis* seedling, E/M.

'Maui' (sport of 'Kona') White. Large heavy anemone form with rippled guard petals. Average bushy growth. USA, 1975. J, M

'Midnight' Black-red. Medium semi-double to anemone form. Vigorous compact upright growth. USA, 1963. J, M

'Mine-no-yuki' ('Snow-white Doves') White. Small semi-double to loose peony form. Can be trained as a standard on a single stem without grafting. J, E

Above: 'Miss Tulare'

Below: 'Mrs D.W. Davis'

'Minimint' White heavily striped pink. Small formal double with high bud centre. Slow bushy growth. USA, 1970. Seedling of 'Donation', M

'Miss Tulare' Bright red to rose red. Large to very large rose-form double to full peony form. Vigorous upright growth. USA, 1945. R, E/M

'Mona Jury' Apricot-pink. Large peony form. Average open growth. USA, 1976. W, E/L

'Moshio' ('Flame') Deep red. Medium semi-double. Upright compact growth. Japan to Australia. J, M

'Moutancha' ('Mudan Cha') Bright pink veined white and striped white on inner petals. Large to very large formal double with wavy, crinkled, crepe-like petals. Average growth. China to USA, 1948. R, L

'Mrs D. W. Davis' Blush-pink. Very large semi-double. Vigorous compact upright growth. USA, 1954. J, M

'Nonie Haydon'

'Myrtifolia Chinesa' Bright pink with lighter pink in centre. Medium rose-form double. Portugal, 1959. J, M

'Navajo' Rose-red fading to white in centre. Medium semi-double. S, E

'Nicky Crisp' Pale lavender-pink. Large semi-double. Slow compact growth. New Zealand, 1980. *C. pitardii* seedling, E/L

'Night Rider' Very dark black-red. Small semi-double. Average compact upright growth. New Zealand, 1985. J, M/L

'Nonie Haydon' Pink. Large peony form. Average growth. New Zealand, 1981. *C. pitardii* seedling, M/L

'Nuccio's Gem' White. Medium to large formal double. Vigorous compact upright growth. USA, 1970. J, E/M

'Nuccio's Jewel' White, washed and shaded orchid pink. Medium full peony form. Slow bushy growth. USA, 1977. J, M

'Nuccio's Pearl' White shaded gently with orchid-pink at edges. Medium formal double. Vigorous compact upright growth.

USA, 1977. J, M

'Nurumi-gata' White shaded pink. Medium to large single with cupped form. Very fragrant. Good hedger. Dense habit, strong growing. *C. oliefera,* E

'Nymph' Pale pink flushed ivory. Miniature semi double. Vigorous spreading growth. Fragrant. New Zealand, 1982. *C. lutchuensis* hybrid, E/L

'Onetia Holland' White. Large to very large loose peony form. Average compact growth. USA, 1954. J, M/L

'Overture' Bright red. Very large semi-double with upright petals. Vigorous compact upright growth. Australia, 1971. R, M

'Pagoda' ('Robert Fortune') Deep scarlet. Large deep formal to rose-form double. Compact growth. China to England, 1857. R, M

'Phyl Doak' Large to very large semi-

'Nuccio's Jewel'

'Plantation Pink'

double. Compact upright growth. New Zealand, 1958. R, E/L

'Pink Cascade' Pale pink. Miniature single with six petals. Weeping growth. New Zealand, 1965. W, M

'Pirates' Gold' Dark red. Large semi-double to loose peony form. Average spreading growth. USA, 1969. J, M/L

'Plantation Pink' Pink. Medium to large single. Good hedger. One of the toughest sasanquas and fast-growing. S, E

'Polar Bear' Chalk white. Large semi-double of hose-in-hose style. Australia, 1957. J, M

'Prudence' Rich pink. Miniature semi-double. Dwarf upright growth. New Zealand, 1971. *C. pitardii* seedling, M

'Purple Gown' Dark purple-red with pinstripes of white to wine red. Large to very large formal double to peony form with wavy petals. Compact growth. China to USA, 1948. R, M

'Pagoda'

'Purple Gown'

'Red Red Rose'

'Queen Diana'/'Diana's Charm' Pink shading to pale pink in outer petals. Medium formal double. Vigorous open growth. New Zealand, 1985. E/L

'Queenslander' Silvery pink. Large rose-form double. Light green foliage. Very strong growth. S, E

'Quintessence' White with yellow anthers and white filaments. Miniature single. Fragrant. Slow spreading growth. New Zealand, 1985. *C. lutchuensis* hybrid, E/M

'Red Red Rose' Bright red. Medium to large formal double with high centre like a rose. Vigorous bushy upright growth. USA, 1969. J, M

'Reigyoku' Orange-red. Small single. Compact growth. Medium. Glossy green foliage with blotch of pink in centre when young, turning to light yellow at maturity.

Japan to USA, 1975. J, M

'Robert Fortune' *see* 'Pagoda'.

'Rosiflora Cascade' Very pale pink. Miniature single. Vigorous weeping open growth. Pronounced cascading habit. (Quite a different appearance from *C. rosiflora.*)

'Sakuraba Tsubaki' Pale pink with fringed petals. Semi-double. The delicately fringed leaves resemble those of a flowering cherry. Japan, 1867. J, M

'Samantha' China pink. Very large semi-double to loose peony form with upright petals. Australia, 1967. R, E/M

'San Dimas' Dark red. Medium to large semi-double with irregular petals. Average compact growth. USA, 1971. J, E/M

'San Marino' Dark red. Large semi-double with heavy textured petals. Average spreading upright growth. USA, 1975. R, M

'Sawada's Dream' White with one-third outer petals shaded delicate flesh-pink. Medium formal double. Average growth. USA, 1958. J, E/M

'Scented Gem' Fuchsia pink with white petaloids. Miniature semi-double. Fragrant. Open upright growth. USA, 1983. E/M

'Scentsation' Silvery pink. Medium to large peony form. Fragrant. Average compact upright growth. USA, 1967. J, M

'Setsugekka' ('Fluted White') White. Medium to large semi-double. Good hedger with pale flowers which don't mark in rain. S, E

'Shishi Gashira' Red. Medium semi-double to double. *C. hiemalis* hybrid, E

'Shot Silk' Brilliant spinel pink. Large semi-double with loose wavy petals. Vigorous growth. China to USA, 1948. R, E

'Showa-no-Sakae' Soft pink, occasionally marbled white. Small to medium semi-double to rose-form double. Good for espalier work. *C. hiemalis* hybrid, E

'Showa Supreme' Soft pink. Medium peony form. *C. hiemalis* hybrid, E

'San Dimas'

'Shishi Gashira'

'Show Girl' Pink. Large to very large semi-double to peony form. Vigorous open upright growth. USA, 1965. R, M

'Sir Victor Davies' Outer petals tone from cardinal red to an old rose with deeper violet veining. Petals pale smoky lavender in the centre. Rose form to peony form. Bushy compact growth. New Zealand, 1990s. J, E/M

'Snippet' Soft pink to almost white centre petals and light-pink outer petals. Small semi-double with long, narrow, notched petals. Dwarf compact growth. New Zealand, 1971. *C. pitardii* seedling, M

'Snowdrop' White edged pink. Miniature single. Open upright growth. Weeping habit. Australia, 1979. E/L

'Sparkling Burgundy' Ruby rose with a sheen of lavender. Small to medium peony form with intermingled stamens and petaloids. Can be trained as a standard on a single stem without grafting. S, E

'Spencer's Pink' Light pink with bright golden-yellow stamens. Large single with wavy petals. Low spreading growth. Australia, 1940. J, E

'Spring Festival' Pink fading to light pink in centre. Miniature rose-form double. Narrow upright growth. USA, 1975. *C. cuspidata* seedling, M/L

'Spring Mist' Blush-pink. Miniature semi-double. Average spreading growth. USA, 1982. *Lutchuensis hybrid*, E/M

'Sugar Babe' Dark pink to red. Miniature compact formal double. Slow growth. USA, 1959. J, M

'Sugar Dream' Pink. Medium anemone form. Average open upright growth. New Zealand, 1984. J, E

'Sunsong' Soft pink. Large formal double.

Average growth. New Zealand, 1980.
Seedling of 'Elegant Beauty', E/L
'Swan Lake' Snow white. Large rose-form
double to loose peony form. Vigorous compact upright growth. Australia, 1968. J, M
'Sweet Emily Kate' Light pink shading to
pale pink in centre. Medium full peony
form. Slow pendulous growth. Fragrant.
Australia, 1987. J, M/L

'Takanini' Deep purplish-red. Small to
medium semi-double anemone form.
Vigorous upright growth. New Zealand,
1989. J, E/L
'Tali Queen' Turkey-red to deep pink.
Very large semi-double with irregular petals,
very large and heavily textured on the
outer ones and wavy on the inner, inter-
spersed with clusters of stamens. Average
upright growth. China to USA, 1948. M
'Tammia' White with pink centre and

Above: 'Sweet Emily Kate'

Below: 'Tali Queen'

'Terrell Weaver'

'The Czar'

border. Miniature to small formal double with incurved geometric petals. Average compact upright growth. USA, 1971. J, M/L

'Tanya' Deep rose pink. Small single. Compact. Has been used successfully as a groundcover and dense hedger. S, E

'Terrell Weaver' Flame to dark red. Large semi-double to loose peony form with thick fluted and twisted petals. Vigorous spreading upright growth. USA, 1974. R, M

'The Czar' Light crimson. Large semi-double. Slow sturdy growth. Australia, 1913. J, M

'The Elf' Light pink. Medium to large semi-double. Slow compact dwarf growth. USA, 1984. J, M

'Tiffany' Light orchid-pink to deeper pink at edge. Large to very large loose peony to anemone form. Vigorous upright growth. USA, 1962. J, M

'Tinker Bell' White striped pink and rose-red. Small anemone form. Vigorous upright growth. USA, 1958. J, E/M

'Tinsie' *see* 'Bokuhan'

'Tiny Princess' White shaded delicate pink. Miniature semi-double to peony form with loose petals and small petaloids. Slow growth. USA, 1961. E/M

'Tiny Star' Soft pink. Miniature semi-

double. Open upright growth. Rather weeping habit. New Zealand, 1978. E/M

'Tiptoe' Silvery pink deepening to cherry-pink at edge. Medium semi-double. Compact upright growth. Australia, 1965. M

'Tom Durrant' Crimson. Large peony form. Medium bushy upright growth. New Zealand, 1966. R, M/L

'Tom Knudsen' Dark red with darker veining. Medium to large formal to rose-form double to full peony form. Vigorous compact upright growth. USA, 1965. J, E/M

'Tiny Princess'

'**Tomorrow Park Hill**' (sport of 'Tomorrow') Light soft pink generally deepening towards the edge with some white variegation. USA, 1964. J, E/M

'**Tootsie**' Chalk white. Miniature formal double, sometimes in the form of a five-pointed star. Slow open spreading growth. USA, 1967. J, M/L

'**Tregrehan**' Apricot-pink. Medium semi-double to rose form double. Vigorous upright growth. England, 1972. W, E/M

'**Twilight**' Light blush-pink. Medium to large formal double. Vigorous compact upright growth. USA, 1964. J, M

'**Un-ryu**' Deep pink. Small single. Average upright growth. Unusual zigzag growth pattern on each leaf. Japan, 1967. J, M

'**Valentine Day**' Salmon pink. Large to very large formal double with rosebud centre. Vigorous upright growth. USA, 1969. R, M

'**Valley Knudsen**' Deep orchid pink. Large to very large semi-double to loose peony form. Vigorous compact upright growth. USA, 1958. R, M/L

'**Ville de Nantes**' (sport of 'Donckelarii') Dark red blotched white. Medium to large semi-double with upright fimbriated petals. France, 1910. J, M/L

Above: 'Valentine Day'

Below: 'Wilamina'

Below: 'Twilight'

'Virginia Franco Rosea' (sport of 'Virginia Franco') Rich rose fading to flesh tone at petal margins. Smallish formal double with centre petals often lightly fringed or feathered. Australia, 1875. J, M

'Waterlily' Lavender-tinted bright pink. Medium formal double. Vigorous compact upright growth. New Zealand, 1967. W, E/M

'White Doves Benten' White. Small semi-double to loose peony form. The 'Bentens' are noted for their brilliantly coloured green leaves, evenly edged with creamy yellow, white or pale green. E

'Wildfire'

'Wilamina' Clear soft pink with darker pink centre and white-tipped edge. Small formal double with incurved petals. Average compact growth. USA, 1951. J, M

'Wilber Foss' Brilliant pinkish-red. Large full peony form. Vigorous upright growth. New Zealand, 1971. W, E/L

'Wildfire' Orange-red. Medium semi-double. Vigorous upright growth. USA, 1963. J, E/M

'William Hertrich' Deep cherry-red. Very large semi-double with heavy irregular petals, the large outer ones somewhat reflexed and the inner ones are smaller, loosely arranged and upright, with some folded and intermixed with stamens. Vigorous bushy growth. USA, 1962. R, M

'Yuletide'

'Wirlinga Belle' Soft pink. Medium single. Average open growth. Australia, 1973. *C. rosiflora* hybrid, E/M

'Wirlinga Gem' Pale pink, deepening at petal edge. Miniature single. Spreading pendulous growth. Australia, 1981. *C. rosiflora* hybrid, E

'Wirlinga Princess' Pale pink fading to white at centre, with deep pink under petals. Average open spreading growth. Has pinkish-bronze spring foliage. Australia, 1977. *C. rosiflora* hybrid, M

'Woodford Harrison' Red veined deep rose. Very large semi-double. Vigorous spreading upright growth. USA, 1980. R, M/L

'Yours Truly' (sport of 'Lady Vansittart') Pink streaked deep pink and bordered white. Medium. Has very attractive shiny holly-like leaves which makes it a good plant for container growing. USA, 1949. J, M

'Yuletide' Orange-red. Small single. Compact. Good for topiary, container growing also a low-growing hedger and espalier subject. S, E

'Zambo' Crimson, veined darker red with some purple overtones. Medium formal double. Australia, 1874. J, M

BIBLIOGRAPHY

Anon. *Azaleas, Rhododendrons, Camellias.* Sunset Books, 1982.

Bielski, Val (ed). *Growing Better Camellias in the 1990s.* New Zealand Camellia Society (Inc.), 1991.

Brownless, Peter (ed). *Camellias, Azaleas and Rhododendrons.* Murray Publishers Ltd, publication date unknown.

Bulletins of the New Zealand Camellia Society, 1974–1995.

Chang, Hung Ta and Bruce Bartholemew. *Camellias.* Timber Press, 1984.

Durrant, Tom. *The Camellia Story.* Heinemann Publishers, 1982.

Edgar, Logan A. *Camellias: The Complete Guide.* The Crowood Press, 1991.

The Garden (various issues). Journal of the Royal Horticultural Society.

Gonos, Arthur A. (ed). *Camellia Nomenclature.* The Southern California Camellia Society Inc., 1993.

Harrison, C.R. and R.E. *Know Your Trees and Shrubs.* A.H. and A.W. Reed, 1966.

Haydon, Neville. Catologues of Camellia Haven Specialist Camellia Nursery, Auckland, New Zealand.

Hay, Roy and Patricia M. Synge. *The Dictionary of Garden Plants.* Ebury Press and Michael Joseph, 1969.

Hume, H. Harold. *Azaleas and Camellias* (revised edition). The Macmillan Company, 1963.

Leach, David. *Rhododendrons of the World.* Charles Scribner's Sons, 1961.

Macaboy, Stirling. *The Colour Dictionary of Camellias.* Landsdowne Press, 1981.

Mathews, Julian (ed). *The New Zealand Gardener* (various issues). INL, 1944–1995.

Palmer, Stanley J. Palmer's *Manual of Trees, Shrubs, and Climbers.* Lancewood, 1990.

Philips, Roger and Martyn Rix. *Shrubs.* Pan Books, Ltd., 1989.

Reader's Digest. *The Reader's Digest Gardener's Encyclopedia of Plants and Flowers.* 1991.

Rolfe, Jim. *Gardening with Camellias: A New Zealand Guide.* Godwit Press, 1992.

Royal Horticultural Society. *Rhododendrons with Magnolias and Camellias* (various issues). Published by The Royal Horticultural Society, London.

Savell, Bob and Stan Andrews. *Growing Camellias in Australia and New Zealand.* Kangaroo Press, 1982.

Simpson, A.G.W. *The Colourful World of Camellias.* Rigby Publishers, 1983.

Sparnon, Norman and E.G. Waterhouse. *The Magic of Camellias.* Ure Smith Pty, Sydney, 1968.

Trehane, David. *Camellias.* The Royal Horticultural Society Wisley Handbook. Cassell, 1991.

Tresseder, Neil and Edward Hyams. *Growing Camellias.* Thomas Nelson & Sons Ltd., 1975.

PICTURE CREDITS

Thank you to the following photographers for the use of their photographs.

INDEX